Springer Tracts in Modern Physics
Volume 120

Editor: G. Höhler
Associate Editor: E. A. Niekisch

Editorial Board:
S. Flügge H. Haken J. Hamilton
W. Paul J. Treusch

Springer Tracts in Modern Physics

* denotes a volume which contains a Classified Index starting from Volume 36

A. Nagl V. Devanathan
H. Überall

Nuclear Pion Photoproduction

With 53 Figures

Springer-Verlag Berlin Heidelberg GmbH

Professor Dr. Anton Nagl
Department of Physics, The Catholic University of America,
Washington, D. C., 20046, USA
and
Pailen-Johnson Associates, Vienna, VA 22182, USA

Professor Dr. Varadarajan Devanathan
Department of Nuclear Physics, University of Madras, Madras 600025, India
and
Crystal Growth Centre, Anna University, Madras 600025, India

Professor Dr. Herbert Überall
Department of Physics, The Catholic University of America,
Washington, D. C., 20046, USA

Manuscripts for publication should be addressed to:
Gerhard Höhler
Institut für Theoretische Kernphysik der Universität Karlsruhe, Postfach 6980,
W-7500 Karlsruhe 1, Fed. Rep. of Germany

*Proofs and all correspondence concerning papers in the process of publication
should be addressed to:*
Ernst A. Niekisch
Haubourdinstraße 6, W-5170 Jülich1, Fed. Rep. of Germany

ISBN 978-3-662-15023-8   ISBN 978-3-540-46066-4 (eBook)
DOI 10.1007/978-3-540-46066-4

© Springer-Verlag Berlin Heidelberg 1991
Originally published by Springer-Verlag Berlin Heidelberg New York in 1991.
Softcover reprint of the hardcover 1st edition 1991

Typesetting: Springer TEX in-house system
57/3140-543210 – Printed on acid-free paper

Preface

Nuclear Pion Photoproduction has become a tool of investigation for (i) the finer details of the elementary photopion production amplitude from free nucleons, (ii) the pion-nucleus optical potential, and (iii) the nuclear structure of the target nucleus. The nuclear cross section involves various bilinear combinations of the different terms in the elementary amplitude and hence serves as an analyzer for the structure of the elementary amplitude. The produced pion undergoes strong interaction with the residual nucleus, and hence the nuclear photopion cross section depends sensitively on the pion-nucleus optical potential. The nuclear structure of the target nucleus plays a dominant role in determining the photopion cross section, and the charged pion photoproduction process with a $T = 0$ target nucleus selectively excites the $T = 1$ final states and hence offers us a powerful tool for investigating $T = 1$ isobar analogue states. In this monograph, all these aspects are considered in detail. In the near future, the experimental study of pion photoproduction will receive a new impetus with the improvement in the technology of pion spectrometers and the commissioning of 100% duty cycle electron accelerators at energies beyond the (3,3) resonance, and it is expected that the investigation of nuclear pion photoproduction will break new ground, and will realize its full potential as a powerful tool of nuclear research. This monograph should serve as an introductory guide as well as a reference manual for the graduate students and research scientists working in this important area of physics.

In the course of writing this book, the authors had the benefit of discussions with numerous original contributors to the subject of nuclear photopion production; particular mention may be made of A. M. Bernstein, R. A. Eramzhyan, N. Freed, V. Girija, K. Shoda, F. Tabakin, L. Tiator and L. E. Wright. Thanks are due to R. Prasad and Reyna Tosta for preparing the manuscript with meticulous care and to S. Karthiyayini for proofreading. The authors acknowledge with thanks the keen interest of Prof. G. Höhler in the publication of this monograph. This work has been supported by the National Science Foundation and one of us (V. D.) acknowledges with thanks the financial support from the Council of Scientific and Industrial Research (India), the award of travel grants by the National Science Foundation and the hospitality at various centers during the preparation of the manuscript.

Washington, D.C. and Madras,   *A. Nagl*
February 1991   *V. Devanathan*
                                                              H. Überall

Contents

1. Introduction

The nuclear pion photoproduction reaction exhibits a number of unique features that make it a rewarding field of study [1.1–14] and a promising source of information on pion-nucleus interaction and nuclear structure. However, even though experimental and theoretical activity in this field started soon after the discovery of the pion and has never ceased ever since, it was a relatively stagnant area of research until recently. The major reasons for this were experimental difficulties and a lack of overlap between the experimentally and theoretically accessible areas of the field.

The low beam intensity and/or low duty cycles of the accelerators available in the 1950's and 1960's limited nuclear pion photoproduction experiments essentially to inclusive reactions in which only the final pion is observed. This meant that the final nucleus could be in a very large number of states, including those which lead to the emission of one or more nucleons. Theoretical predictions, on the other hand, are more likely to be reliable when the final state is well defined [1.15–17]. Only with the commissioning of the high duty cycle, high intensity machines, starting in the mid-seventies, and through the recent advances in detection and data processing equipment, has it been possible to perform with sufficient accuracy pion photoproduction experiments in which transitions to individual nuclear states can be reliably observed. This development has led to a sharp increase in both experimental and theoretical activity in the field, and it has opened the prospect of exploiting the particular characteristics of pion photoproduction, thus making finally accessible the possible benefits that a thorough study of this reaction may offer.

As a source for nuclear structure information, pion photoproduction competes with a variety of other intermediate energy reactions, such as inelastic electron scattering, neutrino reactions, pion scattering, radiative pion capture, and muon capture [1.18–20]. This is due to the fact that all these reactions proceed via similar nuclear transition operators. Studying all of these various reactions does, however, not simply lead to a large amount of redundant information. Rather, it generally leads to complementary information.

Muon capture can provide information about transitions which are hard to study otherwise, but it suffers from the restriction that the momentum transfer is fixed. Electron scattering is free from this restriction, and moreover, it can induce both $\Delta T = 0$ and $\Delta T = 1$ transitions within the same nucleus. Charged pion photoproduction on the other hand connects only to the isobaric analogues of the states reached by the $\Delta T = 1$ transitions in electron scattering. It therefore allows to selectively investigate the isovector transitions. In electron scattering these are mixed up with the isoscalar transitions, and it is sometimes difficult to disentangle them.

Pion photoproduction has, along with electron scattering, the advantage that the momentum transfer to the nucleus is not fixed as it is in muon capture or radiative pion capture, but that it is variable over a wide range, thus allowing to probe a considerable region of the form factors for the various nuclear transitions.

Charged pion photoproduction has in addition the advantage that it proceeds predominantly via spin flip, unlike electron scattering, muon capture, and inelastic pion scattering which excite transitions with and without spin flip. It therefore provides a tool for determining the degree of spin flip contained in the transition to a given level, or to separate out the spin flip components in an unresolved complex of closely spaced transitions.

The interaction of pions with nuclei has thus far been studied mainly through elastic pion scattering and pionic atoms and more recently also through inelastic pion scattering. Here too, pion photoproduction provides an additional investigative tool, complementing the application of the other reactions.

The main advantage of pion photoproduction over elastic pion scattering as a probe of pion-nuclear interaction is that photoproduction provides a much more stringent test for pion wave functions, and hence for the pion-nuclear optical potential, than elastic scattering. The latter reaction is sensitive only to the asymptotic phase shifts. This causes a large ambiguity with respect to the pion wave functions in the nuclear interior, since all wave functions leading to the same phase shifts appear equivalent. Pion photoproduction on the other hand (as is the case with inelastic pion scattering), is very sensitive to the precise nature of the wave functions, particularly in the region in which the nuclear transition densities peak. Since the transition radii for the transitions to the different excited levels vary over a considerable region, it is in principle possible to selectively probe the pion-nucleus interaction within different regions of the nucleus.

The fact that in pion photoproduction, the pions can, in principle, be produced anywhere in the nuclear interior, since the photon can penetrate the whole nucleus essentially unattenuated, makes this reaction a potentially superior probe for pion-nuclear interactions — except for the deep nuclear interior (where the nuclear transition densities tend to be small) and near the (3,3)-resonance region (where the nucleus appears essentially as a black sphere, and where other probes also fail). At energies sufficiently below the (3,3)-resonance, however, where the pion-nuclear interaction is relatively weak, i.e., where the nucleus is relatively transparent to pions, pion photoproduction promises to be an effective tool of pion-nuclear physics.

With pionic atoms, the pion-nuclear interaction can only be studied at one energy ($T_\pi = 0$ MeV). Pion scattering experiments, on the other hand, can only be used for pion energies larger than about 30 MeV. The lower energy limit is imposed by experimental difficulties, stemming from the short distances the pions travel before decaying. In pion photoproduction experiments using spectrometers, a similar problem exists, although to a lesser extent. There are, however, other experimental techniques to measure photoproduction cross sections available which do allow to extend the energy range all the way down to threshold, thereby completely filling the energy gap between pionic atom and pion scattering experiments.

Moreover, in charged pion photoproduction, the pion interacts not with the target nucleus but with the residual nucleus which is usually an unstable isotope with $N \neq Z$. Hence this reaction is extremely suited to test the pion-nucleus optical potential in such an unstable nucleus and study its isospin dependence.

As the preceding discussion indicates, pion photoproduction has the potential of becoming a source of useful information in various parts of nuclear physics. In order to gain access to this source, a large number of experiments have been initiated, promising to yield rapidly increasing amounts of accurate pion photoproduction data. In particular, there has been a growing number of experiments in which transitions to individual nuclear states have been resolved and differential cross sections measured in recent years [1.21]. Correspondingly, there has been a spurt of activity on the theoretical side investigating in detail the various inputs that go into the calculation of photopion cross sections. The purpose of this review is to outline the development, assess the progress and indicate the future prospects of the study of photopion reactions.

This review deals with both the charged and neutral pion photoproduction from nuclei. The charged pion photoproduction takes place incoherently since in the π^+ production, a proton is converted into a neutron and in the π^- production, a neutron is converted into a proton; thereby the final nucleus that is produced in the reaction is different from the initial nucleus. In the case of neutral pion photoproduction, the pion production can take place coherently when the final nucleus is left in the same state as the initial nucleus or incoherently when the nucleus makes a transition to an excited state. It is much easier to detect the charged pions rather than the neutral pions and hence the charged pion photoproduction has received much greater attention. So, the major part of this review is devoted to the study of charged pion photoproduction save Chap. 8 which is devoted entirely to the neutral pion photoproduction.

In Chap. 2, a general survey is made outlining the various phases of development of the subject of nuclear pion photoproduction. Chapters 3–6 deal with the various ingredients that go into the calculation of nuclear photopion cross sections. The elementary amplitude, the construction of nuclear transition amplitudes using the impulse approximation, the final state interaction of the outgoing pion with the residual nucleus and the influence of nuclear wave functions/transition densities are discussed in sequence in successive chapters. In Chap. 7, the present status of the charged pion photoproduction is reviewed and the need for extending the photopion study beyond the (3,3) resonance region is stressed. In Chap. 8, a detailed study of the coherent π^0 photoproduction is made. In Chap. 9, the prospects of the study of polarization phenomena, electroproduction of pions in double coincidence experiments and photoproduction of more exotic mesons such as K and η are briefly discussed.

2. General Survey
of Photopion Nuclear Physics

The development of the subject of nuclear pion photoproduction can be broadly divided into three phases. In the first phase (1947–56), the gross features of the cross sections of photoproduction of pions from hydrogen and various nuclei have been studied. The second phase (1957–76) starts with the rigorous formulation of the elementary photopion production amplitude by Chew et al. using the dispersion theory. In this period, the study on nuclear targets has become more precise by measurement of exclusive total cross sections by observing the radioactivity of the low-lying final nuclear states that are stable against nucleon emission but decay by β-emission and to which the nuclear transition takes place. The third phase (from 1977 onwards) commences with the successful measurement of differential cross sections by Shoda et al. using pion spectrometers. It is during this period that major advances have been made, thus making the nuclear pion photoproduction a powerful tool to investigate the pionic interactions with nuclei and nuclear structure. In Sect. 2.2, different experimental techniques that are used in charged pion spectroscopy are also reviewed. In Sect. 2.3, an outline of the theoretical framework is given for calculating exclusive cross sections.

2.1 Historical Overview

2.1.1 The First Phase (1947–1956)

Interest in pion photoproduction commenced even before the discovery of the pion in 1947 since it was in cosmic rays that the pion events were looked into, ever since Yukawa postulated the meson as the mediator of nuclear force. For a detailed account of photopion reactions during the first phase, the reader is referred to the review article of *Bellamy* [2.1], the book of *Bethe* and *de Hoffmann* [2.2], and other relevant references [2.3–14].

2.1.2 The Second Phase (1957–1976)

A major theoretical advance during this period was the application of dispersion theory to pion photoproduction, in which Watson's theorem played an important part. Based on the cut-off model, which they had already successfully applied to pion scattering, *Chew* et al. (hereafter referred to as CGLN) [2.15] used dispersion relations to obtain a general amplitude for photoproduction of pions from nucleons. Even though a number of simplifying assumptions had to be made, the cross sections calculated with this amplitude compared well with the available data on reactions $\gamma + p \rightarrow \pi^+ + n$ and $\gamma + p \rightarrow \pi^0 + p$. Applications to nuclear photoproduction were made soon after by *Devanathan* and *Ramachandran* [2.16–19] to study charged and neutral pion production from deuterons and later from heavier nuclei.

The first detailed claculations of nuclear pion photoproduction cross sections were attempted by *Laing* and *Moorhouse* [1.15]. Despite the fact that crude approximations were made for the elementary amplitudes and the pion nucleus interactions, as well as for the nuclear transition probabilities, this calculation is important since it was the first attempt to calculate nuclear pion photoproduction cross sections under the assumption that the nucleus undergoes a transition to a discrete final state.

This theoretical advance essentially coincided with an important experimental innovation, due to *Hughes* and *March* [2.20]. They introduced the activity method which allowed for the first time to investigate experimentally transitions to individual final nuclear states with pion photoproduction. Up to that time photoproduction experiments depended on the detection of mesons by determining their charges and energies. Such measurements do not distinguish between reactions in which the meson is the only particle emitted and those in which meson emission is accompanied by the emission of one or more nucleons. If instead of the produced pion, the residual nucleus is observed for some characteristic activity, it is possible to investigate separately reactions in which the nucleus undergoes a transition to individual states stable against particle emission.

The reactions studied by *Hughes* and *March* [2.20] was $^{11}B(\gamma, \pi^-)^{11}C$, where the residual nucleus was identified by observing its positron activity, which has a half life of 20 minutes. Good agreement with theoretical results of *Laing* and *Morehouse* [1.15] was found if the surface production model was assumed to be valid, whereas strong disagreement was found if volume production was assumed (then the predictions were almost an order of magnitude too high).

Soon, a large number of activity experiments were carried out, following the pioneering effort of Hughes and March. Notable among the activity experiments were those of *Dyal* and *Hummel* [2.21] on ^{11}B, of *March* and *Walker* [2.22] on ^{60}Ni, of *Meyer* et al. [2.23] on ^{16}O and ^{27}Al, and of *Nydall* and *Forkman* [2.24] on ^{11}B, ^{27}Al and ^{51}V.

A major advance on the theoretical side were the investigations by *Devanathan* and his collaborators [2.16–19,25–28]. Starting in 1961, this group put the claculations of nuclear pion photoproduction cross sections on a more solid theoretical foundation. Their formalism was based on the impulse approximation, using the CGLN-amplitude for the photoproduction operator, and harmonic oscillator wavefunctions for the nuclear states. Final state interactions between the pion and the nucleus were not considered explicitly. These effects, in particular absorption, were simulated in a phenomenological way by the surface production model, in which it is assumed that pions produced inside the nucleus are reabsorbed, and that therefore the contributions to the cross section come only from the production of pions in the peripheral regions of the nucleus. Here, for the first time, transitions to discrete nuclear states were considered in detailed microscopic shell model calculations.

In 1968, *Saunders* [2.29] undertook a more complete calculation in nuclear pion photoproduction. He included for the first time explicitly final state interactions of the pion with the residual nucleus, using an optical model described by the Kisslinger potential with free pion nucleon scattering parameters.

Another significant theoretical contribution was made by *Berends* et al. (hereafter referred to as BDW) [2.30], who in 1967 used dispersion theory to predict the elementary pion photoproduction multipole amplitudes, basically along the same lines as CGLN [2.15], but carried out in much more elaborate detail.

Some attempts were made to develop a theory of nuclear pion photoproduction in which the nuclei were treated as elementary particles [2.31]. These formalisms lead to simple and elegant expressions. However, they are practically useful only in cases where the vector and axial vector form factors, in terms of which the results are expressed, can be deduced from other sources. Also the method works only if the final nuclear state is a ground state.

By the end of 1970, the photoproduction of pions from free nucleons had been studied extensively in experiments, and it was well understood theoretically. By comparison, investigations on the photoproduction of mesons from complex nuclei were lagging far behind. Theoretically some progress had been made, particularly by Devanathan et al. and Saunders, but experimental information was still scarce and of poor quality. With the exception of half a dozen or so activity experiments, essentially all of the data available up to that point had been obtained in experiments (using bremsstrahlung produced photon beams) in which the photoproduced pions were detected by a counter telescope behind an energy-selecting magnet, i.e., they involved reactions and thus yielded no information on the final nuclear states.

Even the activity experiments did not yield many fruitful results, since they suffered from poor statistics and were usually carried out on nuclei with many discrete final states, thus making detailed theoretical interpretations very difficult.

Theoretical investigations, on the other hand, still suffered from an inadequate treatment of final state interactions (which was largely due to incomplete knowledge of pion-nucleus interactions) and from a lack of accurate nuclear wavefunctions.

Around 1970, the process of nuclear pion photoproduction was beginning to be recognized as a promising tool for studying specific aspects of nuclear structure. *Überall* and his co-workers [2.32,33] used the generalized Goldhaber–Teller Model as a nuclear model to calculate photoproduction cross sections on ^{16}O with the excitation of giant resonance spin-flip states of various multipolarities. In 1973 [2.33], the Helm model was introduced into pion photoproduction calculations as a nuclear model, thus allowing to make use, in a convenient way, of the accurate knowledge of the nuclear form factors gained from the inelastic electron scattering data which had become available.

During the first half of the decade starting from 1970 several other calculations on charged pion photoproduction were carried out, e.g. [2.34–36], but they all were restricted to taking only the $\sigma \cdot \varepsilon$ term into account and/or they neglected final state interactions.

2.1.3 The Third Phase (From 1977 Onwards)

After the new electron linacs went into operation around 1977 and the appearance of accurate experimental data was imminent, interest in carrying out more accurate theoretical studies gained momentum. An additional incentive was provided by the rapidly increasing knowledge concerning the pion-nuclear optical potential [2.37]. This made it possible to take final state interactions in photoproduction into account in a realistic way.

The resurgence of interest in the photopion research at the beginning of this period is evidenced by the conduct of the first international conference on Photopion Nuclear Physics in 1978 and for the status of this subject at that time, the reader is referred to the proceedings of this conference [1.5]. Subsequently, the subject was reviewed at many conferences including the two that were held in Madras [1.6,7].

This period saw also the emergence of a large number of theory groups and notable contributions were made by *Nagl* and *Überall* [2.38], *Singham* and *Tabakin* [2.39], *Decarlo* and *Freed* [2.40], *Girija* and *Devanathan* [2.41], *Tiator* and *Wright* [2.42], *Eramzhyan* and his collaborators [2.43] and *Mukhopadhyay* and his collaborators [2.44]. At the beginning of this period, *Blomqvist* and *Laget* [2.45] derived an expression for the elementary photopion production amplitude which is more suited for investigating the nuclear process. The non-local effects were investigated by *Toker* and *Tabakin* [2.46] and the medium effects by *Dytman* and *Tabakin* [2.47] and they are expected to show up in nuclear transitions where the dominant Kroll–Ruderman term $\sigma \cdot \varepsilon$ is suppressed. The nuclear photopion calculations are usually carried out in coordinate space formalism. The non-local effects can be more easily incorporated in momentum space formalism. Such a program was undertaken by *Tiator* and *Wright* [2.42] and it can be considered as a major advance in the theory of nuclear pion photoproduction.

On the experimental side, a major development was the emergence of the high duty-cycle, high intensity electron linacs. Their availability has made it possible for the first time to carry out pion photoproduction experiments in which transitions to individual levels could be investigated in some detail. The second significant development was the appearance of magnetic pion spectrometers, providing a new high resolution spectroscopic tool in photoproduction physics. So far magnetic spectrometers have been installed at Tohoku, Bates, Saskatchewan and Mainz.

The new linacs allowed experiments with established techniques (activity experiments) with much improved statistics to be carried out, and they enabled experimenters to introduce new methods and tools (such as pion spectrometers) which would have been of little use in connection with the older accelerators.

Besides the Saclay program of $(\gamma, \pi N)$ studies on few-nucleon systems, which is notable for its investigations on the rescattering and meson-exchange effects in quasielastic photoproduction, the experiments involving transitions to discrete nuclear levels were the most important charged pion photoproduction experiments in this period. Three different types of techniques were used in these experiments: the $\pi^+ \mu^+ e^+$-method, based on the decay chain of the pion, the activity method, introduced by Hughes and March in 1957, and the measurements of differential cross sections with pion spectrometers. These methods will be discussed in some more detail in the next section.

The first $\pi \mu e$-experiment, carried out by the Saclay–Louvain collaboration [2.48] on the reaction $^6Li(\gamma, \pi^+)^6He_{g.s.}$, is significant since it was shown for the first time that precise photoproduction experiments were possible with accuracies on the few percent level. Other experiments carried out in the meantime, using this method, were (γ, π^+) on D, which has been studied at Bates [2.49] and at Saclay [2.50], and $^{12}C(\gamma, \pi^+)^{12}B$ and $^{16}O(\gamma, \pi^+)^{16}N$, both done at Bates [2.51,52].

The first of the activity experiments taking advantage of the improved statistics afforded by the new linacs was the investigation of the reaction $^{12}C(\gamma, \pi^-)^{12}N$ at Bates. Originally (1976) this experiment was carried out only in the threshold region [2.53], but in the meantime it has been extended up to the (3,3)-resonance region [2.54]. The first spectrometer experiments were done at Tohoku (1977) by *Shoda* et al. [1.21] and it was the first measurement of differential cross sections for an exclusive photopion production process, thereby permitting a stringent comparison of the theory with the experiment. Soon afterwards (1978), measurements were made at Bates [2.55] with better resolution and improved statistics, but limited to one angle (90°).

With these developments charged pion photoproduction finally seems to have reached a stage, not only where calculations and experiments can be made on the same reaction, but also where theory and experiment are both beginning to be accurate enough so that they can be reliably checked against each other. Large amounts of high quality data are already available and many more are expected in the near future and it is hoped that they will lead to further refinement in the theoretical investigations.

2.2 Experimental Charged Pion Spectroscopy

Three experimental techniques for studying transitions to specific nuclear states through photoproduction of charged pions are currently in use. They have been briefly referred to in the previous section. It will be useful at this point to provide a short review of their characteristics and to indicate their present and possible future applications.

8

2.2.1 The $\pi^+ \mu^+ e^+$ Method

In this technique, the pion and also the muon into which it decays, are stopped in the target, and the e^+, the muon decay product, is detected. The $2.2\,\mu s$ lifetime of the muon makes it possible to count after the beam pulse, thus considerably reducing the background problem. This method, pioneered by the Saclay–Louvain collaboration [2.48], is limited to positive pion production, since negative muons are captured before they can decay. Because of the Coulomb barrier it is restricted to light nuclei, and pion loss from the targets limits the method effectively to energies below 20 or 30 MeV above threshold. Also, since only the final pion is detected (through the end-product of its decay chain), one actually measures an inclusive reaction, i.e., the sum over all final nuclear states, not only the bound states. Therefore the $\pi\mu$ e-method is useful for spectroscopic purposes only until the onset of quasifree pion production.

2.2.2 The Activity Method

Here the radioactivity of the residual nucleus is detected. This method, already introduced in 1957 by Hughes and March, has the advantage that the measurement sums only over those final states which are stable against particle emission. For this reason, the reactions measured with this technique tend to be very exclusive. Its usefulness as a spectroscopic tool is therefore not limited to the threshold region.

Rao et al. [2.28] has pointed out that in particularly favorable cases the final nucleus has only one stable state, as e.g. in the reaction

$$^{12}\text{C}(\gamma, \pi^-)\,^{12}\text{N} \quad .$$

In cases like this, one can investigate the cross sections for transitions to an isolated single state from threshold up to high energies. The above reaction has been investigated on several occasions [2.53,54,56]. Other favorable cases in which the final nucleus has only one bound state (the ground state) is

$$^{13}\text{C}(\gamma, \pi^-)\,^{13}\text{N} \quad \text{and} \quad ^{14}\text{N}(\gamma, \pi^-)^{14}\text{O} \quad .$$

Cases where between one and a few excited bound states are present, are

$$^{7}\text{Li}(\gamma, \pi^-)\,^{7}\text{Be} \quad , \qquad ^{9}\text{Be}(\gamma, \pi^+)\,^{9}\text{Li} \quad ,$$

$$^{10}\text{B}(\gamma, \pi^-)\,^{10}\text{C} \quad , \qquad ^{11}\text{B}(\gamma, \pi^+)\,^{11}\text{Be} \quad ,$$

$$^{13}\text{C}(\gamma, \pi^+)\,^{13}\text{B} \quad , \qquad ^{16}\text{O}(\gamma, \pi^+)\,^{16}\text{N} \quad .$$

For all these reactions, data are available. The cases just mentioned are the most favorable ones from a theoretical point of view, since the cross sections have to be calculated only for a small number of transitions.

The activity method was applied to many other heavier nuclei ranging from ^{27}A to ^{197}Au, but in these cases a theoretical analysis is generally quite difficult because of the large number of final nuclear states involved.

Due to background problems caused by nonmesic interferences, and the effects of the Coulomb barrier, which keeps the cross sections low for positive pions, so far only two experiments have been carried out in the threshold region: $^{11}B(\gamma, \pi^-)^{11}C$ [2.57] and $^{12}C(\gamma, \pi^-)^{12}N$ [2.53]. In the resonance region, on the other hand, the activity method has been applied to a large variety of nuclei.

2.2.3 Pion Spectrometers

This research tool has added a new dimension to photonuclear physics. With their availability it became possible to measure differential cross sections, and therefore provide a much more sensitive test for theoretical predictions than possible with total cross sections. It also became possible to vary the energy and momentum transfer separately and therefore test various components of the photoproduction process individually (e.g. test either the nuclear form factors or the pion-nucleus interaction, depending on whether the momentum transfer or the energy transfer is varied).

With a monochromatic photon source the observed pion spectrum would consist of a number of discrete lines, each providing the cross section for the transition to an individual state. The use of a bremsstrahlung beam, which is necessary to obtain realistic counting rates, complicates matters considerably. Instead of a discrete set of lines, a superposition of staggered yield curves is obtained, one for each excited state (Fig. 2.1). The individual yield curves are shifted relative to the first one by the excitation energy of the state to which they belong. The weight factor with which a yield curve is present in the composite is a measure of the cross section for the transition to the corresponding state.

Figure 2.1 shows the total pion counting rate plotted vs. pion energy. The range between T_n and T_{n+1} in pion energy in Fig. 2.1 indicates the width of the spectrometer acceptance window, and the total pion yield curve is in fact a composite pieced together from measurements in a number of different such energy ranges. In Fig. 2.1 the total yield curve shows a sharp break every time a new state comes in. In practice, however, because of the finite resolution, the transition is gradual. Hence, if many weakly excited states come in, in rapid

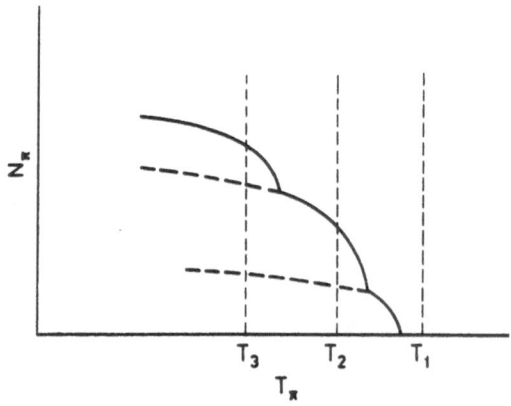

Fig. 2.1. Typical pion spectrometer counting rate as a function of pion energy

succession, it will not be possible to resolve them. On the other hand, if the separation between levels is of the order of 1 MeV and the excitation of these levels is sufficiently strong, then at present a separation can be made with some confidence. It is expected that in the near future it will be possible to resolve levels with a separation of the order of 100 keV.

Several spectrometer data are already available from Tohoku [1.9] and Bates [2.55,58,59] and they provide a stringent test for the theoretical inputs and the investigations of the effects of non-locality and medium modifications of the elementary operator.

2.3 Theoretical Framework

The theoretical study of the photopion production process in nuclei involves the following four essential factors, viz.,

1) the elementary photopion production amplitude for free nucleons,
2) the construction of the nuclear transition amplitude from the free nucleon amplitude,
3) the final state interaction of the outgoing pion with the residual nucleus, and
4) the nuclear structure.

For the elementary amplitude, one has several choices. *Chew* et al. [2.15] and *Berends* et al. [2.30] have deduced the single nucleon amplitude from dispersion theoretical considerations and later *Blomqvist* and *Laget* [2.45] have deduced it from a consideration of the Born diagrams that contribute to the photoproduction process and adding the delta contribution ad hoc. All these elementary amplitudes have been tested and they are found to yield single nucleon cross sections fairly well. The extension to the nuclear problem is usually made by invoking the impulse approximation. In the early calculations [1.15,17], the outgoing pions were represented by plane waves. Since they yielded much larger cross sections, a surface production model was invoked to take into account somewhat the off-shell effects and the final state interactions of the outgoing pion with the residual nucleus. It is to be stressed that the surface production model is purely a phenomenological model. In later works [2.38–41], this model is replaced by a more rigorous approach. The off-shell effects are taken into account by replacing the pion momentum by the gradient operator in the configuration space formalism. The final state interaction of the outgoing pion with the residual nucleus is treated by using distorted pion waves obtained by solving the Klein–Gordon equation with the pion-nucleus optical potential that reproduces the pion-nucleus elastic cross section. In a similar way, the nuclear wave functions should be chosen after carefully testing them with the data obtained from electron scattering, beta decay and muon capture. Alternatively, the nuclear transition densities obtained from the electron scattering data using a phenomenological Helm model [1.18] can be used.

In the following chapters, the aforementioned factors that go into the calculation of cross sections for photoproduction of pions from nuclei are discussed in detail and the theoretical results are compared with the available experimental data. The theoretical inputs are analysed and a study is made on the extent of their validity.

3. Elementary Photopion Production Amplitude

3.1 Introduction

A photon incident on a nucleon couples to the nucleon electromagnetic current, causing the nucleon to radiate mesons if the photon energy is sufficiently high. Up to photon energies of 1 GeV the production of photomesons is dominated by the production of single pions and pion pairs.

The measured cross sections [3.1] shown in Fig. 3.1 [3.2] for the reaction $\gamma p \to n\pi^+$ indicate that photoproduction of single charged pions can be considered as being determined by a number of N* resonances and in addition by processes which provide a smoothly varying background.

The background (indicated in Fig. 3.1 by a dashed line) is due to the Born part of the production amplitude which arises from s- and u-channel nucleon pole diagrams and from t-channel meson exchange processes. In the energy range indicated, four nucleon resonances couple into the photopion amplitude: the P_{33}, P_{11}, D_{13} and S_{11} resonances located at c.m. energies of 1236, 1400, 1520 and 1535 MeV, respectively. Their contributions interfere with the background and with each other. However, as can be seen from Fig. 3.1, photoproduction of single charged pions is dominated by the effect of a single large resonance, the P_{33} ($J = 3/2$, $T = 3/2$) Δ resonance, up to at least 800 MeV photon energy. This observation is underscored by the shape of the solid line which represents the

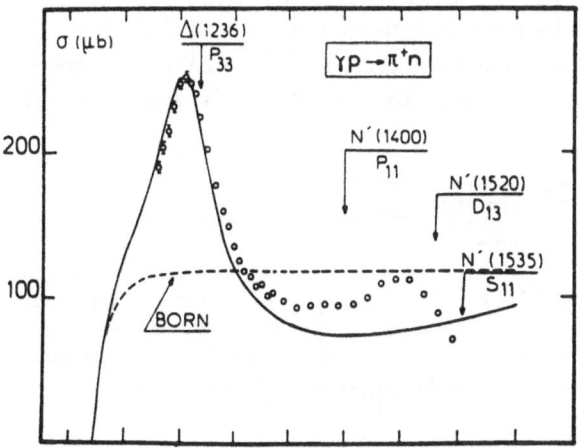

Fig. 3.1. Total cross section for the reaction $\gamma p \to n\pi^+$. Experimental points from [3.1], curves are the prediction of the Blomqvist–Laget model [3.3]. The resonances are labeled as $L_{2I,2J}$ [3.2]

13

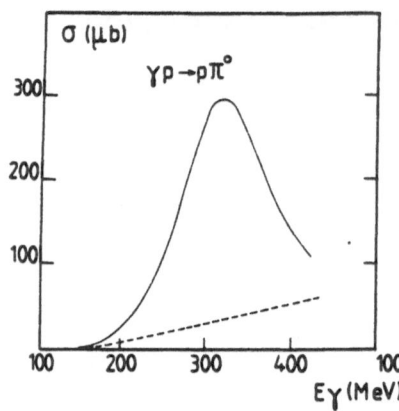

Fig. 3.2. Total cross section of $\gamma p \to p\pi^0$ [3.2]. Broken line: Born term only

calculated cross section, taking only the effects of the background (Born) terms and the Δ resonance into account [3.3].

For neutral pion production a similar picture emerges, except that here the Δ resonance dominates the cross section even more strongly. The solid curve in Fig. 3.2 [3.1] shows the calculated total cross section for $\gamma p \to p\pi^0$, the dashed line the result if only the Born terms are considered [3.3].

Since pions are 0^- mesons they are photoproduced by axial currents. Furthermore, pions are isovector particles. They can thus couple both to isospin 1/2 and isospin 3/2 N^* resonances. These two factors, together with gauge and Lorentz invariance and the unitarity condition, determine the general structure of the production amplitude.

While the threshold for two-pion production is near 300 MeV photon energy, the cross section for the $\gamma N \to N\pi\pi$ process is insignificant up to 400 MeV, thus introducing negligible inelasticities in the resonance phase shifts at least up to that energy. Between threshold and 700 MeV total cross section data [3.4] shown in Fig. 3.3 [3.2] for $\gamma p \to p\pi^+\pi^-$, can be almost completely explained as being due to the creation of a $\pi\Delta$ pair through a contact term (photon, nucleon, pion and Δ interacting at the same point) and a photoelectric term. This reaction is thus also strongly dominated by the Δ resonance, at least up to 700 MeV.

The threshold for η meson production occurs at 709.3 MeV photon energy. As an isoscalar particle, the η can couple only to isospin 1/2 N^* resonances. The

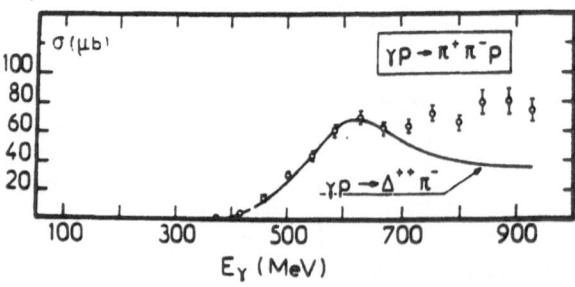

Fig. 3.3. Total cross section for $\gamma p \to p\pi^+\pi^-$. Experimental points are from [3.4]. Curve corresponds to cross section of $\gamma p \to \Delta^{++}\pi^-$ [3.2]

production threshold for kaons is found at $E_\gamma = 911\,\text{MeV}$ (if the target nucleon is changed into a Λ) or $E_\gamma = 1046\,\text{MeV}$ (if the target nucleon is changed into a Σ). Of special interest is K^+ production on protons ($\gamma p \rightarrow K^+ \Lambda^0$ and $\gamma p \rightarrow K^+ \Sigma$) since here no resonances are superimposed on and interfering with the Born (background) amplitude. Since both η mesons and kaons are 0^- mesons like the pion, η and kaon photoproduction amplitudes have the same basic structure as the photopion amplitudes.

Production thresholds for vector mesons are above 1 GeV. While all meson production processes mentioned above have been investigated both theoretically and experimentally (see e.g. [3.2] for references), single pion photoproduction is by far the most studied and best understood.

In the photoproduction of a pion on a free nucleon asymptotically three types of fields are involved, the electromagnetic field A^μ, the pion field ϕ_π^α, and the nucleonic field ψ_N. The electromagnetic field interacts with the charge and the magnetic moment of the nucleon, and also with the meson cloud surrounding the nucleon. Furthermore, since the pion and the nucleon interact strongly, complicated rescattering processes of the outgoing pion on the nucleon take place, leading to multi-particle intermediate states in the s- and u-channel. In addition, all multi-pion and vector meson intermediate states in the t-channel which asymptotically lead to one-pion final states are found to contribute.

The smooth background component of the photopion cross section is found to be dominated by the electric dipole (E_{0+}) amplitude. This amplitude is primarily determined by the seagull diagram, which represents a point interaction among all three asymptotic fields. The seagull graph (along with the photoelectric diagram) accounts for the effect of a virtual pion surrounding a nucleon being converted into a real pion by interacting with the electromagnetic field of the incident photon.

The large contribution of the seagull graph (and the photoelectric term) to the E_{0+} multipole, and thus to the background amplitude, is missing in neutral pion production since only charged pions can couple to the photon and be ejected. Neutral pion production thus has a very small background component.

Interactions of the photon with the nucleon magnetic moment contribute primarily to the M_{1+} multipole. Contributions involving nucleon charges reflect the coupling of the photon to the current due to the motion of the initial or the recoiling final nucleon. They are, however, generally not significant.

The major rescattering mechanisms proceed through the resonant $M_{1+}^{3/2}$ and $E_{1+}^{3/2}$ multipole channels. The importance of the Δ resonance in both charged and neutral pion photoproduction (which is readily apparent in Figs. 3.1 and 3.2, and also in Fig. 3.4, which compares experimental points from [3.5] with the calculations of [3.3] for the $\gamma n \rightarrow p\pi^-$ reaction) is a direct consequence of the fact that the $J = T = 3/2$ channel is resonant and dominant in pion-nucleon scattering. Watson's theorem [3.6] requires that at least one of the multipoles in the photoproduction amplitude proceeding through that channel is also resonant. From discussion of the multipole expansion in Sect. 3.3 it follows that both the electric quadrupole amplitude E_{1+} and the magnetic dipole amplitude M_{1+} contribute to

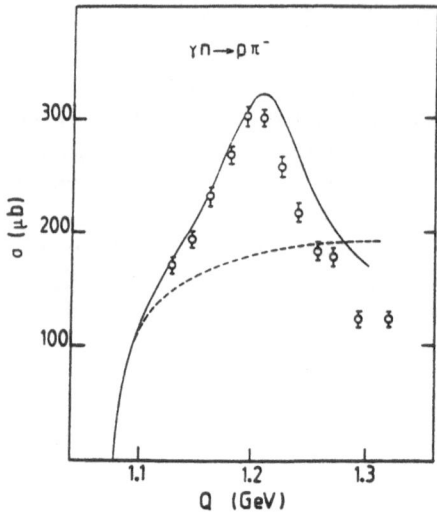

Fig. 3.4. Total cross section of $\gamma n \to p\pi^-$ [3.2]. Experimental points from [3.5]. Broken line: Born term only

the (3,3) channel. According to the quark model, the Δ is obtained from the nucleon by a quark spin flip, implying an $M1$ transition for this resonance. Recently, the quark model has also predicted an $E2$ contribution which is however certain to be small [3.7]. The two most important multipoles in the photoproduction of pions on free nucleons from threshold on past the resonance region are thus the electric dipole amplitude, which dominates the background (Born) component, and the magnetic dipole amplitude which dominates the resonance component of the total amplitude.

Important as these two multipoles are they allow only order of magnitude estimates of experimentally determined cross sections. A number of other multipoles are found to contribute substantially. Moreover, if the elementary photopion operator is to be embedded in a nuclear medium, as e.g., in a DWIA calculation, the relative importance of different multipoles may shift significantly, sometimes strongly enhancing the importance of otherwise less prominent multipole components. This fact underscores the need for a detailed and reliable description of the elementary photopion process.

As a first step it is useful to establish the most general form of the photopion amplitude which is consistent with applicable invariance principles. This subject is addressed in Sect. 3.2. The multipole decomposition of the general amplitude in terms of angular momentum, parity and isospin, which is convenient both for analyzing experiments and for theoretical investigations, is described in Sect. 3.3. For pion photoproduction, unitarity considerations play a prominent role. Some implications are discussed in Sect. 3.4.

Several approaches have been used to derive explicit forms of the photopion amplitude, i.e., to quantitatively describe the dynamics of the photoproduction process on a nucleon. From the perspective of nuclear applications the goal is to find a consistent, physically founded and transparent form of the amplitude which accurately reproduces experimental data, is in agreement with low energy theorems, satisfies unitarity constraints, and which above all can be unambiguously embedded into nuclear calculations. No completely satisfactory solution to this problem has yet been found.

The simplest approach is given by the Born approximation, which is limited to the evaluation of the nucleon and pion pole contributions of the total amplitude. The Born approximation can be expected to provide a good description of low energy photo-pion production. Since the s-wave pion-nucleon interaction is weak, there is little rescattering of the pion after it is created. This is indeed the case for charged pion production. For neutral pion production the two isospin components making up the Born amplitude accidentally cancel to a high degree, resulting in a Born term so small that it is exceeded by the resonant amplitude already at 2 MeV above threshold. However, in either case the Born amplitude becomes exact in the limit $E_\pi \to 0$. Even though the Born amplitude is not sufficiently accurate for most applications, it is of considerable interest since most other approaches build on the Born approximation in one way or another. The Born approximation is discussed in some detail in Sect. 3.5.

The first successful attempt to go beyond the Born approximation in a systematic way was based on the use of dispersion relations. This approach, which effectively iterates the Born approximation to all orders, was pioneered by *Chew*, *Goldberger*, *Low*, and *Nambu* [3.7], subsequently referred to as CGLN. Using this approach, in which unitarity and analyticity are strictly imposed, Chew et al. derived an explicit expression for the photopion amplitude which was based on the assumption that the Δ resonance dominates the dispersion integrals.

Höhler et al. [3.9] evaluated the CGLN equations in more detail, stressed the importance of recoil effects and demonstrated that for photon energies in the Δ resonance region the CGLN theory compared favorably with experimental data. Especially the third reference [3.9] presents a large number of graphical fits of CGLN results to the data.

Berends, *Donnachie* and *Weaver* [3.10], in a quite elaborate effort, have used the fixed-t dispersion relations to evaluate photopion amplitudes for photon lab energies of up to 500 MeV. They used fully relativistic projections for all multipoles. Whereas the CGLN analysis was limited to s- and p-wave amplitudes (with the exception of the photoelectric term which was kept to all orders), Berends et al. obtained all multipole amplitudes up to $l \le 3$.

The analysis of photopion reactions within the framework of dispersion relations has led to a good understanding of the basic mechanisms involved, and good agreement with experiment has been achieved. This approach is, however, not very suitable for energies much beyond the Δ resonance. Since the results obtained with this approach are formulated in the pion-nucleon c.m. frame, they cannot be easily and unambiguously adapted for nuclear calculations. The dispersion-theoretical approach is discussed in Sect. 3.6.

An efficient and physically transparent way to define the photopion amplitude is the effective Lagrangian approach of *Blomqvist* and *Laget* [3.3]. Here the Born amplitude is supplemented by an s-channel Δ exchange diagram with empirically determined coupling constants. The Lagrangians, which describe the coupling among photons, mesons, nucleons and baryonic resonances, are chosen in such a way that they satisfy the constraints imposed by the low energy theorems and the PCAC hypothesis. A non-relativistic reduction of the resulting amplitude to order $(p/M)^2$ is applied. The transition operator obtained is Lorentz and gauge invariant and valid in any reference frame. This feature has made the operator a popular choice for use in nuclear problems. The Blomqvist–Laget operator as well as some attempts to improve it are discussed in Sect. 3.7.

3.2 The Invariant Amplitude

This section contains a discussion of the general form of the transition amplitude for the elementary pion photoproduction process. Following the convention of *Bjorken* and *Drell* [3.11], the S-matrix for the process is written as

$$S_{fi} = \delta_{fi} - \frac{i}{(2\pi)^2} \delta^4 (p_f + q - k - p_i) \left(\frac{M^2}{4 E_f E_\pi E_\gamma E_i} \right)^{1/2} T_{fi} \qquad (3.1)$$

where k, q, p_i, p_f and E_γ, E_π, E_i, E_f are, respectively, the 4-momenta and the energies of the photon, the pion and the initial and final nucleon, and M is the nucleon mass. The T-matrix element T_{fi} is given as

$$T_{fi} = \varepsilon_\mu J_\mu^{fi} \qquad (3.2)$$

where ε_μ is the photon polarization vector, and

$$J_\mu^{fi} = \overline{u}_f(p_f, s_f) J_\mu u_i(p_i, s_i) \qquad (3.3)$$

is the matrix element of the nucleon electromagnetic current J_μ, and u_i and u_f are the initial and final nucleon Dirac spinors. The current J_μ is a Lorentz covariant pseudo-four vector composed of the particle 4-momenta and the Dirac matrices. The most general form for the transition operator

$$T = \varepsilon_\mu J_\mu \qquad (3.4)$$

is obtained as a linear combination

$$T = \sum_{i=1}^{N} A_i M_i \qquad (3.5)$$

of all independent Lorentz invariants M_i which can be formed by combining the polarization vector ε, the Dirac matrices γ, and the particle momenta.

Because of the properties the γ matrices only terms of the form enter:

$\gamma_5 \, f_1(Q_1, Q_2) \slashed{\varepsilon} \slashed{k}$

$\gamma_5 \, f_2(k, Q_1, Q_2) \slashed{\varepsilon}$

$\gamma_5 \, f_3(\varepsilon, Q_1, Q_2) \slashed{k}$

$\gamma_5 \, f_4(k, \varepsilon, Q_1, Q_2)$,

where for the two momenta Q_1 and Q_2 any convenient combination may be chosen, such as $Q_1 = q$, $Q_2 = (p_i + p_f)/2$ or $Q_1 = p_i$, $Q_2 = p_f$. The slashed 4-vectors $\not\varepsilon$ or $\not k$ represent the scalar products $\gamma \cdot \varepsilon$ and $k \cdot \varepsilon$, respectively. Gauge invariance limits the number of linearly independent Lorentz invariants M_i to four [3.8]. The set chosen by *Chew* et al. [3.8] and *Berends* et al. [3.10] (corresponding to the choice $Q_1 = q$, $Q_2 = (p_i + p_f)/2$) is

$$M_1 = i\,\gamma_5 \not\varepsilon \not k$$

$$M_2 = 2i\,\gamma_5 (P \cdot \varepsilon\, q \cdot k - P \cdot k\, q \cdot \varepsilon) \qquad (3.6)$$

$$M_3 = \gamma_5(\not\varepsilon\, q \cdot k - \not k\, q \cdot \varepsilon)$$

$$M_4 = 2\gamma_5(\not\varepsilon\, P \cdot k - \not k\, P \cdot \varepsilon - iM\not\varepsilon\not k) \quad .$$

Other equivalent forms can be obtained by linear combination by using momentum conservation. Using (3.3–5), the transition matrix element (3.2) is written in terms of the invariant operators M_i and the scalar amplitudes A_i as

$$T_{fi} = \bar{u}_f(p_f, s_f)\, \varepsilon_\mu J_\mu\, u_i(p_i, s_i) = \sum_j A_j\, \bar{u}_f(p_f, s_f)\, M_j\, u_i(p_i, s_i) \quad . \qquad (3.7)$$

Expressing the Dirac spinors and the γ matrices in terms of Pauli spinors and matrices, the operators M_i sandwiched between the Dirac spinors yield the explicit form

$$T_{fi} = \sum_j A_j(s, t)\, M_j^{fi} \qquad (3.8)$$

with

$$\frac{2M}{\sqrt{e_i e_f}}\, M_1^{fi} = \left\langle f \left| ik\,\sigma \cdot \hat{\varepsilon} + \hat{\varepsilon} \cdot \left(\frac{p_f}{e_f} - \frac{p_i}{e_i} \right) \times k \right. \right.$$

$$+ i\left[\sigma \cdot k \left(\frac{p_i}{e_i} + \frac{p_f}{e_f} \right) \cdot \hat{\varepsilon} - \sigma \cdot \hat{\varepsilon} \left(\frac{p_i}{e_i} + \frac{p_f}{e_f} \right) \cdot k \right] \qquad (3.9a)$$

$$+ \frac{ik}{e_i e_f}\, [p_i \cdot p_f \sigma \cdot \hat{\varepsilon} - p_i \cdot \hat{\varepsilon}\sigma \cdot p_f - \sigma \cdot p_i p_f \cdot \hat{\varepsilon}]$$

$$\left. + \frac{k}{e_i e_f}\, \hat{\varepsilon} \cdot (p_i \times p_f) \left| i \right\rangle \right.$$

$$\frac{2M}{\sqrt{e_i e_f}}\, M_2^{fi} = \left\langle f \left| 2i\,\sigma \cdot \left(\frac{p_f}{e_f} - \frac{p_i}{e_i} \right) \right. \right.$$

$$\left. [P \cdot \hat{\varepsilon}(q \cdot k - E_q k) - q \cdot \hat{\varepsilon}(P \cdot k - kE_p)] \left| i \right\rangle \right. \qquad (3.9b)$$

$$\frac{2M}{\sqrt{e_i e_f}} M_3^{fi} = \left\langle f \left| -i\sigma \cdot \hat{\varepsilon}(q \cdot k - E_q k) + i\sigma \cdot k q \cdot \hat{\varepsilon} \right. \right.$$

$$-ik\sigma \cdot \left(\frac{p_i}{e_i} + \frac{p_f}{e_f}\right) q \cdot \hat{\varepsilon}$$

$$+\frac{E_q k}{e_i e_f} \left[\sigma \cdot p_f p_i \cdot \hat{\varepsilon} + \sigma \cdot p_i p_f \cdot \hat{\varepsilon} - p_i \cdot p_f \sigma \cdot \hat{\varepsilon}\right]$$

$$+\frac{\hat{\varepsilon} \cdot p_i \times p_f}{e_i e_f}(q \cdot k - E_q k) - q \cdot \hat{\varepsilon} \frac{k \cdot p_i \times p_f}{e_i e_f} \qquad (3.9c)$$

$$+\frac{i}{e_i e_f} \left[\sigma \cdot p_f(p_i \cdot k q \cdot \hat{\varepsilon} - p_i \cdot \hat{\varepsilon} q \cdot k)\right.$$

$$+\sigma \cdot p_i(p_f \cdot k q \cdot \hat{\varepsilon} - p_f \cdot \hat{\varepsilon} q \cdot k)$$

$$\left. \left. +p_i \cdot p_f(\sigma \cdot \hat{\varepsilon} q \cdot k - \sigma \cdot k q \cdot \hat{\varepsilon})\right] \right| i \right\rangle$$

$$\frac{2M}{\sqrt{e_i e_f}} M_4^{fi} = \left\langle f \left| -2i\sigma \cdot \hat{\varepsilon}(q \cdot k - E_q k) + 2i\sigma \cdot k P \cdot \hat{\varepsilon} \right. \right.$$

$$+2ik\sigma \cdot \left(\frac{k}{e_i} + \frac{q}{e_f}\right) P \cdot \hat{\varepsilon}$$

$$+\frac{2}{e_i e_f} \hat{\varepsilon} \cdot p_i \times p_f (P \cdot k - E_p k)$$

$$-\frac{2}{e_i e_f} k \cdot (p_i \times p_f) P \cdot \hat{\varepsilon} \qquad (3.9d)$$

$$-\frac{2i}{e_i e_f} \left[\sigma \cdot p_f p_i \cdot \hat{\varepsilon} + \sigma \cdot p_i p_f \cdot \hat{\varepsilon} - p_i \cdot p_f \sigma \cdot \hat{\varepsilon}\right]$$

$$\times (P \cdot k - E_p k) \left. \right| i \right\rangle - 2M \frac{2M}{\sqrt{e_i e_f}} M_1^{fi} \quad .$$

In (3.9) the quantities E_p, e_i, e_f and E_q stand for $E_p = (E_i + E_f)/2$, $e_i = E_i + M$, $e_f = E_f + M$, and the pion energy, respectively. Expressions (3.9) are valid for any coordinate frame and for arbitrary energies. They are separately gauge invariant even if this is no longer obvious from the expanded forms. The forms (3.9) are however not unique since instead of using the set (3.6) one could have started with a different set of gauge and Lorentz invariant matrices M_i, e.g., the set

$$M_1 = i\gamma_5 \slashed{\varepsilon} \slashed{k}$$

$$M_2 = 2i\gamma_5 (\varepsilon \cdot p_i k \cdot p_f - \varepsilon \cdot p_f k \cdot p_i) \qquad (3.10)$$

$$M_3 = \gamma_5 (\slashed{\varepsilon} k \cdot p_i - \slashed{k} \varepsilon \cdot p_i)$$

$$M_4 = \gamma_5 (\slashed{\varepsilon} k \cdot p_f - \slashed{k} \varepsilon \cdot p_f)$$

obtained from (3.6) by using momentum conservation to eliminate q and by linear combination of M_1, M_3, and M_4. The dynamics of the photoproduction process is contained in the four scalar amplitudes A_i which depend only on the coupling constants and the Mandelstam variables

$$s = (p_i + k)^2$$

$$t = (k - q)^2 \tag{3.11}$$

$$u = (p_i - q)^2 \quad .$$

The analytic properties of these amplitudes are determined by the possible intermediate states in the s-, t-, and u-channels. In the s- and u-channel there are poles at M^2 associated with single nucleon intermediate states, corresponding to the direct and crossed nucleon Born terms (Figs. 3.5a and c). The pole in the t-channel at $t = m^2$ (m = pion mass) is associated with the single-pion exchange diagram, the pion Born term (Fig.3.5e). Branch cuts in the s- and u-channels starting at $(M + m)^2$ are associated with nucleon-plus-one-pion intermediate states, corresponding, respectively, to the rescattering and crossed rescattering terms (Figs. 3.5b and d). In the t-channel there are cuts at $t > 4m^2$ and $t > 9m^2$, corresponding to two-pion and three-pion exchange diagrams (Figs. 3.5f and g). Additional cuts in each channel are associated with higher order diagrams.

Equation (3.8) can be rewritten as

$$T_{fi} = \left\langle f \left| \sum_{i=1}^{N} F_i \theta_i \right| i \right\rangle \tag{3.12}$$

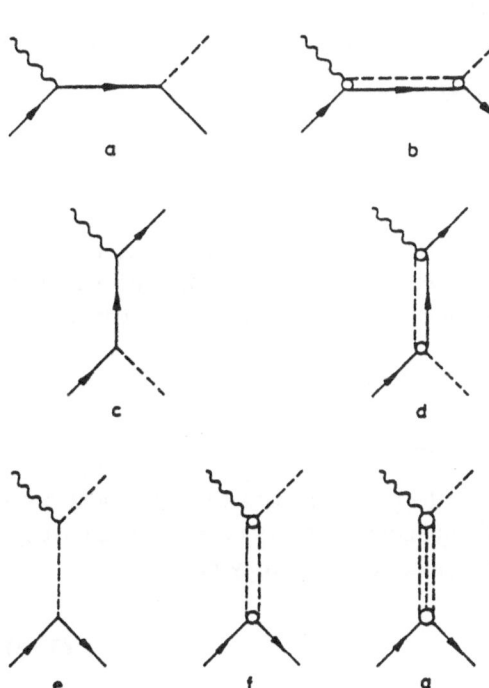

Fig. 3.5a–g. Diagramatic structure of the photopion amplitude [3.10]. (a) s-channel pole diagram (direct nucleon Born term), (b) s-channel nucleon-plus-one-pion intermediate state (rescattering term), (c) u-channel pole diagram (crossed nucleon Born term), (d) u-channel nucleon-plus-one-pion intermediate state (crossed rescattering term) (e) t-channel pole diagram (pion Born term), (f) two pion exchange diagram (ϱ-exchange term), (g) three pion exchange diagram (ω-exchange term)

where the operators θ_i are scalar and vector products composed of the Pauli spinor σ, the photon polarization vector $\hat{\varepsilon}$ and the four particle momenta p_i, p_f, q, and k. The amplitudes F_i are found as linear combinations of the invariant amplitudes A_i. In the most general form, (3.12) consists of a total of $N = 31$ terms. When using momentum conservation or when expressing (3.12) in a specific coordinate system in which the nucleon momenta are explicitly related to q and k, the number N of terms obtained for (3.12) reduces considerably.

Additional simplifications are obtained if only terms up to some power of p/M, p being any nucleon momentum, are kept. For the pion-nucleon c.m. frame, which is appropriate for pion photoproduction on a free nucleon, (3.12) reduces to 4 terms:

$$
\tilde{T}_{fi} = \frac{4\pi W}{M} \left\langle f \left| i\sigma \cdot \hat{\varepsilon} \, F_1 + \frac{\sigma \cdot q \, \sigma \cdot k \times \hat{\varepsilon}}{|q| \, |k|} F_2 \right. \right.
$$
$$
\left. \left. + i \frac{\sigma \cdot k \, q \cdot \hat{\varepsilon}}{|q| \, |k|} F_3 + i \frac{\sigma \cdot q \, q \cdot \hat{\varepsilon}}{q^2} F_4 \right| i \right\rangle \quad . \tag{3.13}
$$

Explicit relations between the coefficients F_i in (3.13) and the invariant amplitudes $A_i(s,t)$ in (3.8) are given in [3.10], which in the form used in [3.8] are

$$
\frac{4\pi}{W-M} \frac{2W}{\sqrt{e_i e_f}} F_1 = A_1 - \frac{k \cdot q}{W-M} (A_3 - A_4) + (W - M) A_4
$$

$$
\frac{4\pi}{W+M} \frac{2W}{qk} \sqrt{e_i e_f} \, F_2 = -A_1 - \frac{k \cdot q}{W+M} (A_3 - A_4) + (W + M) A_4
$$

$$
\frac{4\pi}{W+M} \frac{2W}{qk} \sqrt{\frac{e_i}{e_f}} \, F_3 = (W - M) A_2 + A_3 - A_4
$$

$$
\frac{4\pi}{W-M} \frac{2W}{q^2} \sqrt{\frac{e_f}{e_i}} \, F_4 = -(W + M) A_2 + A_3 - A_4 \quad .
$$

$$(3.14)$$

In the definition of the amplitudes F_i in (3.13) the factor $4\pi W/M$, W being the total energy in c.m. frame, has been factored out [3.10]. With this definition the cross section for the transition from an initial γ-nucleon state i to a final pion-nucleon state f is given by

$$
\frac{d\sigma}{d\Omega} = \frac{q}{k} \sum \left| \frac{M}{4\pi W} \tilde{T}_{fi} \right|^2
$$

where the sum represents summing over the photon polarization states and summing and averaging over the magnetic quantum numbers of the nucleon states. At threshold $F_2 = F_3 = F_4 = 0$, and the cross section simply becomes

$$
\left. \frac{d\sigma}{d\Omega} \right|_{thr} = \frac{q}{k} |F_1|^2 \quad . \tag{3.15}
$$

Equations (3.8 and 9) represent the most general photopion amplitude valid at all energies and in any coordinate frame. As they stand these equations are not very useful, however, unless the invariant amplitudes $A_j(s,t)$ are explicitly known as a function of s and t over some region of interest.

Theoretical analyses of the photopion amplitude are generally carried out in the pion-nucleon c.m. frame. Using the inverse of relations (3.14), the amplitudes A_j may be evaluated for specific values of s and t, from which subsequently the photopion amplitude in an arbitrary frame may be obtained. This procedure is of some interest when the elementary transition operator is applied to pion photoproduction on nuclei. For the Born component explicit expressions for the invariant amplitudes A_i may be given. They are discussed in Sect. 3.5.

3.3 Partial Wave Decomposition

Expanding the photon wave function into vector spherical harmonics Y_{JL} one obtains for each eigenvalue J of the total angular momentum three states with different L values, classified as electric ($L = J \pm 1$) and magnetic ($L = J$) multipoles. An electric 2^J-pole has parity $(-)^J$, a magnetic 2^J-pole has parity $-(-)^J$. If the photon multipoles are combined with the spin of the nucleon, the angular momentum configurations given in Table 3.1 result.

The columns j and P specify the total angular momentum and parity of the possible configurations of the initial and final states. The values given in the column l indicate the angular momentum values allowed for the pion. Because

Table 3.1. Decomposition of pion photoproduction amplitude into multipole components

Photon J	Photon Multipole	Total j	P	Pion l	Multipole Amplitude
1	$E1$	1/2	−	0	E_{0+}
		3/2	−	2	E_{2-}
	$M1$	1/2	+	1	M_{1-}
		3/2	+	1	M_{1+}
2	$E2$	3/2	+	1	E_{1+}
		5/2	+	3	E_{2-}
	$M2$	3/2	−	2	M_{2-}
		5/2	−	2	M_{2+}
3	$E3$	5/2	−	2	E_{2+}
		7/2	−	4	E_{4-}
	$M3$	5/2	+	3	M_{3-}
		7/2	+	3	M_{3+}
4	$E4$	7/2	+	3	E_{3+}
		9/2	+	5	E_{5-}
	$M4$	7/2	−	4	E_{4-}
		9/2	−	4	E_{4+}
etc.					

of the negative intrinsic parity of the pion only those values of l are allowed for which $-(-)^l = P$. For each pair of values of j and P there are two possible channels through which the reaction can proceed: an electric and a magnetic multipole transition. The amplitudes for these channels are commonly assigned the symbols given in the last column. The first subscript indicates the angular momentum l of the pion, the second one the total angular momentum of the intermediate state; $\pm$ stands for $j = l \pm 1/2$. Thus for each angular momentum value of the pion there are in general four different amplitudes: $E_{l\pm}$ and $M_{l\pm}$.

Electromagnetic interactions, even though they do not conserve isospin, nevertheless behave in all well-defined manner under isospin transformations. The interaction Hamiltonian consists of a term H_s, transforming like an isotopic scalar, and a term H_{v3} which transforms like the third component of an isotopic vector (and, if terms quadratic in A_μ are kept, possibly also a term transforming like an isotopic tensor). In pion photoproduction the initial state has isospin $T = 1/2$, the final state can be $T = 1/2$ or $T = 3/2$. For H_s the selection rule $\Delta T = 0$ holds, whereas for H_{v3} transitions involving $\Delta T = 0, \pm 1$ are allowed. Consequently, three isospin amplitudes are necessary and also sufficient to describe the photoproduction of pions on a nucleon:

$$\left\langle \frac{1}{2}, T_3 \,\middle|\, T_s \,\middle|\, \frac{1}{2}, T_3 \right\rangle$$

$$\left\langle \frac{1}{2}, T_3 \,\middle|\, T_{v3} \,\middle|\, \frac{1}{2}, T_3 \right\rangle$$

$$\left\langle \frac{3}{2}, T_3 \,\middle|\, T_{v3} \,\middle|\, \frac{1}{2}, T_3 \right\rangle \quad .$$

Since physical final states are obtained as a linear combination of the isospin eigenstates, i.e.

$$|\pi N\rangle = a \left| \frac{1}{2}, T_3 \right\rangle + b \left| \frac{3}{2}, T_3 \right\rangle \quad ,$$

the photoproduction amplitude in a physical channel is obtained as

$$\langle \pi N' | T | \gamma N \rangle = \sqrt{3}\, a \left\langle \frac{1}{2} \,\|\, T_s \,\|\, \frac{1}{2} \right\rangle + \frac{2}{\sqrt{3}}\, a T_3 \left\langle \frac{1}{2} \,\|\, T_v \,\|\, \frac{1}{2} \right\rangle$$

$$= \sqrt{3}\, a A^{(0)} + \frac{2}{\sqrt{3}}\, T_3 a A^{(1)} + \sqrt{\frac{2}{3}}\, b A^{(3)} \qquad (3.16)$$

where the A's are the reduced matrix elements of the isospin eigenamplitudes. Inserting the appropriate coefficients for the various combinations of pions and nucleons in (3.16) one obtains

$$\langle \pi^\pm N^\mp | T | \gamma N^\pm \rangle = \sqrt{2} \left(A^{(0)} \pm \frac{A^{(1)} - A^{(3)}}{3} \right)$$

$$\langle \pi^0 N^\pm | T | \gamma N^\pm \rangle = \pm A^{(0)} + \frac{A^{(1)} + 2A^{(3)}}{3} \qquad (3.17)$$

where $N^{\pm}$ stands for proton and neutron, respectively. These expressions give the decomposition of the physical photoproduction amplitudes into isospin channels of definite transformation properties.

The connection between the isospin decomposition

$$A_j(s,t) = A_j^{(+)} \frac{\{\tau_\alpha, \tau_3\}}{2} + A_j^{(-)} \frac{[\tau_\alpha, \tau_3]}{2} + A_j^{(0)} \tau_\alpha \tag{3.18}$$

of the invariant amplitudes $A_i(s,t)$ (which is, e.g., obtained when evaluating the photopion amplitude in the Lagrangian approach, see Sect. 3.5) and the isospin eigenamplitudes is given by the relations

$$A_j^{(-)} = \frac{A_j^{(1)} - A_j^{(3)}}{3} \quad , \qquad A_j^{(+)} = \frac{A_j^{(1)} + 2A_j^{(3)}}{3} \tag{3.19}$$

with the isoscalar amplitudes $A_i^{(0)}$ having the same meaning in both cases.

The isospin decomposition carries over into the multipole amplitudes. Since the angular momentum decomposition yields 4 different amplitudes per pion partial wave $(E_{l\pm}, M_{l\pm})$, altogether twelve different amplitudes are needed for each l:

$$E_{l\pm}^{(0)}, \quad E_{l\pm}^{(1)}, \quad E_{l\pm}^{(3)}, \quad M_{l\pm}^{(0)}, \quad M_{l\pm}^{(1)}, \quad M_{l\pm}^{(3)} \quad .$$

The connection between the multipole expansion (i.e. essentially the angular momentum representation) and the momentum representation (3.13) of the total photoproduction amplitude, i.e., the relation between the coefficients F_i and the multipole amplitudes $E_{l\pm}$ and $M_{l\pm}$ is accomplished through the use of projection operators [3.8], and is given by

$$F_1 = \sum_{l=0}^{\infty} \left[l M_{l+} + E_{l+} \right] P'_{l+1}(x) + \left[(l+1) M_{l-} + E_{l-} \right] P'_{l-1}(x)$$

$$F_2 = \sum_{l=1}^{\infty} \left[(l+1) M_{l+} + l M_{l-} \right] P'_l(x) \tag{3.20}$$

$$F_3 = \sum_{l=1}^{\infty} \left[E_{l+} - M_{l+} \right] P''_{l+1}(x) + \left[E_{l-} + M_{l-} \right] P''_{l-1}(x)$$

$$F_4 = \sum_{l=2}^{\infty} \left[M_{l+} - E_{l+} - M_{l-} - E_{l-} \right] P''_l(x)$$

in terms of the derivatives of the Legendre polynomials, x being the cosine of the pion c.m. angle.

3.4 Implications of Unitarity

Through unitarity of the S-matrix (implying conservation of flux), photoproduction and radiative pion capture are linked to pion scattering. In addition, both are connected to Compton scattering of photons on nucleons. As a consequence the S-matrices for the processes (γ, γ'), (γ, π), (π, γ), and (π, π') cannot be considered isolated by themselves, but only as part of an enlarged S-matrix:

$$S = \begin{pmatrix} S_{\gamma\gamma} & S_{\gamma\pi} \\ S_{\pi\gamma} & S_{\pi\pi} \end{pmatrix} \quad . \tag{3.21}$$

If the unitarity condition is expressed in terms of the T-matrix, one obtains for each total angular momentum submatrix

$$2 \operatorname{Im} T_{fi}^j = \sum_n T_{fn}^j T_{ni}^{j*} \quad . \tag{3.22}$$

If the initial and final states are taken to be (γ, N) and (π, N), respectively, and if it assumed that of the n possible intermediate states (γN, πN, $\gamma\pi N$, $\pi\pi N$, etc.), only one is important, namely the final state (π, N), one has

$$2 \operatorname{Im} T_{fi}^j = T_{ff}^j T_{fi}^{j*} \quad . \tag{3.23}$$

This result implies that there is no final state photon (which is reasonable because of the smallness of the fine structure constant), and that the energy of the incident photon remains below the two-pion threshold.

If the πN S-matrix is taken to be unitary, which is also a reasonable assumption, since the photoproduction matrix elements are smaller than the pion scattering elements by almost two orders of magnitude, then

$$T_{ff}^j = T_{\pi\pi}^j = \frac{S_{\pi\pi}^j - 1}{i} = \frac{e^{2i\,\delta_j^\pi} - 1}{i} \quad . \tag{3.24}$$

Hence one has

$$\frac{T_{fi}^j - T_{fi}^{j*}}{2i} = \frac{e^{2i\,\delta_j^\pi} - 1}{2i} T_{fi}^{j*} \quad . \tag{3.25}$$

Since one term cancels on each side, one finds

$$T_{fi}^j = |T_{fi}^j| \, e^{i\,\delta_j^\pi} \quad . \tag{3.26}$$

For each value of j there are 12 different input configurations in photoproduction, grouped together according to their orbital angular momentum and isospin values:

$$l = j - \frac{1}{2}, \quad T = \frac{1}{2} : \quad M_{l+}^{(0)}, \; M_{l+}^{(1)}, \; E_{l+}^{(0)}, \; E_{l+}^{(1)}$$

$$l = j - \frac{1}{2}, \quad T = \frac{3}{2} : \quad M_{l+}^{(3)}, \; E_{l+}^{(3)}$$

$$l = j + \frac{1}{2}, \quad T = \frac{1}{2} \quad : \quad M_{l-}^{(0)}, \ M_{l-}^{(1)}, \ E_{l-}^{(0)}, \ E_{l-}^{(1)}$$

$$l = j + \frac{1}{2}, \quad T = \frac{3}{2} \quad : \quad M_{l-}^{(3)} \ E_{l-}^{(3)} \quad .$$

All configurations in a particular group lead to the same final pion-nucleon state, an angular momentum and isospin eigenstate characterized by l and T.

The T-matrix T_{fi}^{j} separates into submatrices T_{fi}^{jlT} since there is no coupling between members of different submatrices. Hence relation (3.26) holds for each submatrix:

$$T_{\mathrm{fi}}^{jlT} = \left| T_{\mathrm{fi}}^{jlT} \right| \mathrm{e}^{\mathrm{i}\,\delta_{jlT}^{\pi}} \quad . \tag{3.27}$$

Therefore, all members of a particular group characterized by (l, T) have the same phase shift, i.e., the amplitudes $M_{l\pm}^{(0)}$, $E_{l\pm}^{(0)}$, $M_{l\pm}^{(1)}$ and $E_{l\pm}^{(1)}$ all have the pion phase shifts $\delta_{l\pm1/2,l,1/2}^{\pi}$, whereas $M_{l\pm}^{(3)}$ and $E_{l\pm}^{(3)}$ have the pion phase shifts $\delta_{l\pm1/2,l,3/2}^{\pi}$. Relation (3.23) applied to photoproduction is known as Watson's theorem [3.16]. This theorem is strictly valid up to the two pion threshold ($E_\gamma \sim$ 300 MeV) and approximately valid as long as the partial wave inelasticities remain small. The first phase shift which shows major inelasticity is δ_{11}. Since the related photoproduction amplitudes are small, the Watson theorem is useful past the (3,3)-resonance region up to photon energies of close to 500 MeV. The theorem has important implications, since as long as it is applicable, the number of parameters required to describe photoproduction is reduced by a factor of two: only the modulus of the amplitudes has to be specified, the phase angles can be taken from pion scattering.

The unitarity condition imposes an important constraint on the construction of photopion amplitudes. In the Born approximation all multipoles can be made unitary by multiplying them with the corresponding Watson phase factor. When extending the photopion amplitude beyond the Born approximation, care has to be taken to maintain unitarity of the total amplitude.

In models which derive photopion multipole amplitudes from dispersion relations, the unitarity constraint can be built in from the start. In phenomenological models, however, in which individual resonances are added empirically to a non-resonant background, the unitarity condition is usually not satisfied, even though each component amplitude (for the background and for each resonance) may be separately unitary. In these cases unitarity has to be explicitly restored, which is accomplished by multiplying the resonance amplitude by some complex factor.

The physical origin of this renormalization factor lies in the interference effects between resonances and background. For the Δ resonance the interference is caused by pions which are produced through the background interaction and then rescatter on the nucleon via the Δ.

3.5 The Born Approximation

The Born amplitude consists of the contribution of all poles to the invariant amplitudes. These pole contributions correspond to single particle intermediate states. The Born approximation thus consists of neglecting all contributions to the photopion amplitude associated with branch cuts. The latter are related to resonances and multiple-particle intermediate states.

If pseudo-scalar (PS) coupling between pions and nucleons is chosen, the Born terms consist of the nucleon exchange in the direct or s-channel, pion exchange in the t-channel, and nucleon exchange in the crossed or u-channel. The resulting amplitude is gauge-invariant. In the pseudo-vector (PV) coupling mode an additional diagram (the seagull or contact term) is needed to achieve gauge invariance. The contact term is obtained when the minimal substitution $q_\mu \rightarrow q_\mu - ie\varepsilon_\mu$ is made in the charged pion PV Lagrangian. The diagrams contributing to the Born amplitude in PS and PV coupling respectively, are indicated in Fig. 3.6.

The derivation of the Born amplitude corresponding to these diagrams is straightforward and unambiguous except for the vertex functions coupling pions and nucleons.

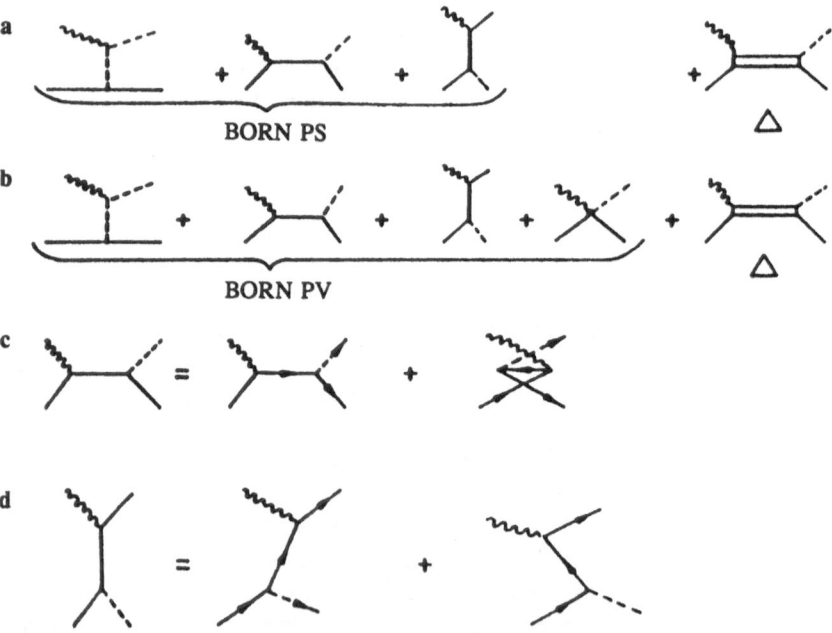

Fig. 3.6a–d. The different diagrams considered in pion photoproduction on nucleons. The wavy line, broken line, and full line represent respectively the photon, pion and nucleon. The Δ (1236) intermediate state is represented by a double line. (a) and (b) are respectively the PS Born terms and PV Born terms plus s-channel Δ (1236) formation. (c) and (d) are time ordered decomposition of direct and crossed nucleon Born terms [3.3]

For all diagrams other than the seagull diagram the reaction matrix elements have the general form [3.3]

$$T_{fi} = -\overline{u}_f(p_f, s_f)\, \Gamma_f \frac{Q}{P}\, \Gamma_i\, u_i(p_i, s_i) \tag{3.28}$$

where u_i and u_f are the initial and final nucleon spinors, Γ_i and Γ_f the vertex operators and Q/P the propagator of the intermediate particle.

For the γNN vertex, the vertex function is

$$\Gamma_{\gamma NN} = e \left[\frac{1+\tau_3}{2}\, \gamma_\mu - \frac{\sigma_{\mu\nu}}{2M}\, k^\nu \left(\frac{\kappa_p + \kappa_n}{2} + \tau_3\, \frac{\kappa_p - \kappa_n}{2}\right)\right] e^\mu u \tag{3.29}$$

where κ_p and κ_n are the anomalous magnetic moments of the proton and neutron.

The first term describes the interaction of the photon with the convection current of the nucleon. In the barycentric system it contributes only through the recoil motion of a final proton. The second term represents the interaction of the photon with the magnetic moment of the nucleon. Since the isovector part dominates over the isoscalar part ($\kappa_p + \kappa_n \ll \kappa_p - \kappa_n$), the photon-nucleon interaction proceeds primarily through the $T = 3/2$ channel.

The $\gamma\pi\pi$ vertex describes the interaction of the electromagnetic field of the photon with the charge of a virtual pion emitted by the nucleon, leading to the ejection of the pion from the vicinity of the nucleon. The vertex factor here is

$$\Gamma_{\gamma\pi\pi} = e_\pi \varepsilon_\mu (q + q')^\mu \tag{3.30}$$

where e_π is the pion charge, and q' and q are the 4-momenta of the intermediate and the final pion.

Using PV coupling between the pion and the nucleon, the πNN vertex function is

$$\Gamma_{\pi NN}^{PV} = \frac{f}{m}\, \overline{u}_f(\slashed{p}_f - \slashed{p}_i)\, \gamma_5 \tau_\alpha u_i \tag{3.31}$$

for the case in which both nucleon legs are external, and

$$\overline{\Gamma}_{\pi NN}^{PV} = \frac{f}{m}\, \slashed{q}\, \gamma_5 \tau_\alpha u \tag{3.32}$$

for a vertex linked to a diagram involving a nucleon as the intermediate particle. Here p_i and p_f are, respectively, the initial and final nucleon momenta and q the pion momentum. The sign of the charge of the pion involved is $-\alpha, \tau_\alpha$ being the nucleon isospin transition operator. For pseudo-scalar coupling the vertex functions are, respectively,

$$\Gamma_{\pi NN}^{PS} = g\, \overline{u}_f \gamma_5 \tau_\alpha u_i \tag{3.33}$$

and

$$\overline{\Gamma}_{\pi NN}^{PS} = g\, \gamma_5 \tau_\alpha u_i \quad , \tag{3.34}$$

g and $f = mg/2M$ being the pseudoscalar and pseudovector coupling constants.

The Seagull Diagram. The seagull term is the result of the point interaction of the two currents

$$j_{\pi N} = \frac{f}{m} \delta_\mu \overline{\psi}_f \gamma_5 \gamma^\mu \tau_\alpha \psi_i \tag{3.35}$$

and

$$j_{\pi\gamma} = e_\pi \left[\partial^\mu \left(A_\mu \phi_\alpha \right) + A_\mu \left(\partial^\mu \phi_\alpha \right) \right] \tag{3.36}$$

where ψ_i, ψ_f are the initial and final nucleon fields and ϕ_α is the field of a pion of charge $e_\pi = -\alpha e$ expressed in terms of the electric charge $e = \sqrt{4\pi/137}$. After quantitizing the fields and keeping only the appropriate creation and annihilation operators, a transition matrix element of

$$T_{fi}^{SG} = e_\pi \tau_\alpha \frac{f}{m} \overline{u}_f \gamma_5 \slashed{\epsilon} u_i \tag{3.37}$$

is obtained. Using the identity

$$e_\pi \chi_t^{f\dagger} \tau_\alpha \chi_t^i = e \chi_t^{f\dagger} \frac{[\tau_\alpha, \tau_3]}{2} \chi_t^i \tag{3.38}$$

where χ_t^i and χ_t^f are the initial and final nucleon isospinors, (3.37) can be written as

$$T_{fi}^{SG} = -\frac{[\tau_\alpha, \tau_3]}{2} \frac{ef}{m} \overline{u}_f \gamma_5 \slashed{\epsilon} u_i \quad . \tag{3.39}$$

The isospin commutator yields $\pm\sqrt{2}$ for $\pi^\pm$ production and vanishes for neutral pion production as required. Evaluated to order $(p/M)^2$, (3.39) yields

$$T_{fi}^{SG} = \frac{ef}{m} \chi_t^{f\dagger} \frac{[\tau_\alpha, \tau_3]}{2} \chi_t^i \langle f | i\boldsymbol{\sigma} \cdot \hat{\boldsymbol{\epsilon}} | i \rangle \quad . \tag{3.40}$$

The seagull diagram provides the dominant contribution to the photopion amplitude for low energy charged pion production. In PS coupling the seagull term does not contribute. In this case the contribution equivalent to that of the seagull term is provided by the antiparticle component of one of the nucleon exchange diagrams, as discussed below.

Expression (3.40) is just the result obtained from the Kroll–Ruderman theorem as representing the exact Hamiltonian in the limit of $E_\pi \to 0$. At the physical threshold $E_\pi = m$ the Kroll–Ruderman result is only approximate. Applying the PCAC hypothesis allows one to improve the approximation. The other diagrams in the Born approximation also contain terms proportional to $\boldsymbol{\sigma} \cdot \hat{\boldsymbol{\epsilon}}$, i.e., they also provide a correction to the simple Kroll–Ruderman result (3.40). However, neither the PCAC result nor the full Born approximation provides the exact result at threshold. This situation is discussed in more detail in the next section.

Since $\varepsilon_0 = 0$, the Kroll–Ruderman term (3.40) can be rewritten, using

$$i\boldsymbol{\sigma} \cdot \hat{\boldsymbol{\epsilon}} = i(\sigma_+ \varepsilon_- + \sigma_- \varepsilon_+) \tag{3.41}$$

and one can then see that it describes the transfer of a projected component $|\Delta m| = 1$ of angular momentum from the photon to the nucleon. If the nucleon spin is initially aligned antiparallel to the photon momentum ($m = -1/2$) then only the first term in (3.41) applies. According to it, an incident photon with right-hand circular polarization will give up its unit of angular momentum to the nucleon, thereby causing the nucleon spin to flip to the state $m = +1/2$. Since in the long wavelength limit ($k_\gamma = 0$), where (3.40) is exact, the total angular momentum of the photon is 1, and since this amount is transferred to the nucleon, the pion is necessarily produced in a relative s-state. Hence for the final state $J = 1/2$, $l = 0$, $P = 1$. Thus the multipole causing the transition is necessarily an electric dipole and the amplitude describing this process in the usual nomenclature is E_{0+}.

The Photoelectric Diagram (Pion Pole Term). The photoelectric diagram is in principle similar to the seagull diagram. In both cases a virtual pion is emitted by the nucleon, and the resulting charge configuration interacts with the electromagnetic field of the incident photon. The basic difference between the two cases is that in the seagull diagram the pion is produced in an s-state ($l = 0$), whereas in the process described by the photoelectric diagram it is emitted in a higher angular momentum state ($l \geq 1$).

For the pion pole term the following transition matrix element is obtained

$$T_{\mathrm{fi}}^{\mathrm{PE}} = - \overline{u}_{\mathrm{f}} g\, \gamma_5 \tau_\alpha u_{\mathrm{i}}\, \frac{1}{t - m^2}\, i\, e_\pi 2q \cdot \varepsilon \quad . \tag{3.42}$$

Using again relation (3.38), (3.42) can be rewritten as

$$T_{\mathrm{fi}}^{\mathrm{PE}} = i\, eg\, \overline{u}_{\mathrm{f}}\, \frac{[\tau_\alpha, \tau_3]}{2}\, \gamma_5\, \frac{2q \cdot \varepsilon}{t - m^2}\, u_{\mathrm{i}} \quad . \tag{3.43}$$

A non-relativistic reduction to order $(p/M)^2$ yields

$$T_{\mathrm{fi}}^{\mathrm{PE}} = i\, \frac{eg}{M}\, \chi_t^{\mathrm{f}\dagger}\, \frac{[\tau_\alpha, \tau_3]}{2}\, \chi_t^{\mathrm{i}}\, \left\langle f\left| \frac{q \cdot \hat{\varepsilon}\, \sigma \cdot (p_{\mathrm{i}} - p_{\mathrm{f}})}{E_q k - q \cdot k} \right| i \right\rangle \quad . \tag{3.44}$$

For the photoelectric diagram the same result is obtained whether PS coupling (as assumed in (3.42)) or PV coupling is used.

The Direct Nucleon Pole Term. Assuming PS coupling, the transition matrix element for nucleon exchange in the s-channel is found to be

$$T_{\mathrm{fi}}^{\mathrm{ND,PS}} = \overline{u}_{\mathrm{f}}\, g\tau_\alpha \gamma_5\, \frac{\not{p}_{\mathrm{i}} + \not{k} + M}{s - M^2}$$

$$e\left[\frac{1 + \tau_3}{2}\, \gamma_\mu - \frac{\sigma_{\mu\nu}}{2M} k^\nu \left(\frac{\kappa_{\mathrm{p}} + \kappa_{\mathrm{n}}}{2} + \tau_3\, \frac{\kappa_{\mathrm{p}} - \kappa_{\mathrm{n}}}{2} \right) \right] \varepsilon^\mu\, u_{\mathrm{i}}. \tag{3.45}$$

This can be reduced to components proportional to $\not{k}\not{\varepsilon}$, $\not{k}$, $\not{\varepsilon}$, and 1:

31

$$T_{fi}^{ND,PS} = ge\,\overline{u}_f \left\{ \tau_\alpha \left(\frac{1+\kappa_p+\kappa_n}{2} + \frac{1+\kappa_p-\kappa_n}{2}\tau_3 \right) \frac{\gamma_5 \slashed{k}\slashed{\varepsilon}}{s-M^2} \right.$$

$$+ \tau_\alpha \left(\frac{\kappa_p+\kappa_n}{2M} + \frac{\kappa_p-\kappa_n}{2M}\tau_3 \right) \frac{\gamma_5}{s-M^2} \left(p_i \cdot k\slashed{\varepsilon} - p_i \cdot \varepsilon\slashed{k} \right)$$

$$\left. - \tau_\alpha \frac{1+\tau_3}{2} \frac{2\gamma_5\, p_i \cdot \varepsilon}{s-M^2} \right\} u_i \ . \tag{3.46}$$

A convenient alternate evaluation of the nucleon pole diagram is obtained if the nucleon propagator is split into a particle and an anti-particle component [3.3], corresponding to the two time-ordered diagrams shown in Fig. 3.6c.

In the first of these diagrams (the particle diagram) the intermediate particle is a nucleon: after absorbing the photon, the nucleon propagates in a virtual state until the pion is emitted. The second diagram (the antiparticle diagram), which is not present in a nonrelativistic theory, assumes as the initial event the spontaneous creation of the pion together with a nucleon-antinucleon pair, e.g. a π^+ and a neutron-antiproton pair. The nucleon (e.g. the neutron) escapes, whereas the antinucleon (e.g. the antiproton) propagates to the point where it annihilates with the incident nucleon (e.g. the proton) while absorbing the incident photon. To order $(p/M)^2$ one obtains for the PS-amplitude:

$$T_{fi}^{ND,PS} = \frac{f}{m} e \left\langle f \left| \chi_t^{f\dagger} \tau_\alpha \left\{ \frac{\boldsymbol{\sigma} \cdot \boldsymbol{q}\, \boldsymbol{\sigma} \cdot \boldsymbol{k} \times \hat{\boldsymbol{\varepsilon}}}{2Mk} \right. \right. \right.$$

$$\times \left(\frac{1+\kappa_p+\kappa_n}{2} + \tau_3 \frac{1+\kappa_p-\kappa_n}{2} \right)$$

$$- i\, \frac{\boldsymbol{\sigma} \cdot \boldsymbol{q}\, p_i \cdot \hat{\boldsymbol{\varepsilon}}}{Mk} \frac{1+\tau_3}{2}$$

$$+ i\boldsymbol{\sigma} \cdot \hat{\boldsymbol{\varepsilon}} \left[-\frac{1+\tau_3}{2} + \frac{k}{2M} \right.$$

$$\left. \left. \left. \times \left(\frac{\kappa_p+\kappa_n}{2} + \tau_3 \frac{\kappa_p-\kappa_n}{2} \right) \right] \right\} \chi_t^i \left| i \right\rangle \ . \tag{3.47}$$

For the PV-amplitude the result is the same as (3.47), except that the factor associated with the third term inside the curly bracket is replaced by a simpler expression

$$-\frac{1+\tau_3}{2} + \frac{k}{2M} \left(\frac{\kappa_p+\kappa_n}{2} + \tau_3 \frac{\kappa_p-\kappa_n}{2} \right) \rightarrow \frac{E_\pi}{2M} \frac{1+\tau_3}{2} \ . \tag{3.48}$$

According to (3.47) and (3.48), the total amplitude for the direct nucleon Born diagram is seen to consist of three terms:

1) The first term is the result that would be obtained in a nonrelativistic theory using the static Chew–Low Hamiltonian for the pion-nucleon vertex. It corresponds to the photon interacting with the magnetic moment of the nucleon leading to a p-wave emission of a pion.

2) The second term is also generated by the first of the time-ordered diagrams in Fig. 3.6c. This term corresponds to a kinematic correction, taking into account the effect of the motion of the initial nucleon, leading in first order to an electric dipole interaction with the charge of the moving nucleon. The isospin factor limits the appearance of this term to the cases where the initial nucleon is a proton. Furthermore, this term vanishes if the amplitude is evaluated in the laboratory system ($p_i = 0$) or in the pion-nucleon barycentric system ($p_i = -k$).

3) The third term in (3.47) is due to the antiparticle component of the time-ordered diagrams. It is therefore absent in a non-relativistic theory. Whereas the non-relativistic treatment leads only to a p-wave emission of the pion, the antiparticle term is proportional to $\sigma \cdot \hat{\varepsilon}$, and hence corresponds to the emission of an s-wave pion. This is also the only term in which the PS- and the PV-theory differ. For PV-coupling the s-wave term is small, for PS-coupling, on the other hand, this term, if the initial nucleon is a proton, provides a large contribution, which in fact essentially compensates for the absence of the seagull term in the PS-theory. If the initial nucleon is a neutron, the s-wave term, however, is small even for PS-coupling.

The Crossed Nucleon Pole Term. For nucleon exchange in the u-channel (corresponding to the third diagram in Fig. 3.6a or b) the transition matrix element, assuming PS coupling, is

$$T_{\text{fi}}^{\text{NC,PS}} = -\bar{u}_f \varepsilon^\mu \left[\frac{1 + \tau_3}{2} \gamma_\mu - \left(\frac{\kappa_p + \kappa_n}{2} + \tau_3 \frac{\kappa_p - \kappa_n}{2} \right) \right]$$
$$\times \sigma_{\mu\nu} \frac{k^\nu}{2M} \frac{\not{p}_f - \not{k} + M}{u - M^2} \gamma_5 \tau_\alpha u_i \ . \tag{3.49}$$

A reduction similar to that of the s-channel term yields

$$T_{\text{fi}}^{\text{NC,PS}} = ge\, \bar{u}_f \left\{ \left(\frac{1 + \kappa_p + \kappa_n}{2} + \frac{1 + \kappa_p - \kappa_n}{2} \tau_3 \right) \tau_\alpha \frac{\gamma_5 \not{k} \not{\varepsilon}}{u - M^2} \right.$$
$$+ \left(\frac{\kappa_p + \kappa_n}{2} + \frac{\kappa_p - \kappa_n}{2} \tau_3 \right) \tau_\alpha \frac{\gamma_5}{u - M^2} \left(p_f \cdot k\not{\varepsilon} - p_f \cdot \varepsilon \not{k} \right)$$
$$+ \left. \frac{1 + \tau_3}{2} \tau_\alpha \frac{\gamma_5}{u - M^2} 2 p_f \cdot \varepsilon \right\} u_i \ . \tag{3.50}$$

For the u-channel nucleon exchange a decomposition of the propagator leads to terms corresponding to the two time-ordered diagrams shown in Fig. 3.6d. The second part of the diagram corresponds again to the antiparticle term specific to the relativistic theory. It describes the creation of a nucleon-antinucleon pair by the incident photon, of which the nucleon escapes and the antinucleon anihilates with the incident nucleon, resulting in the emission of a pion.

For PS coupling, the following non-relativistic form to order $(p/M)^2$ is found:

$$T_{\mathrm{fi}}^{\mathrm{NC,PS}} = \frac{f}{m}\, e \left\langle f \left| \chi_t^{\mathrm{ft}} \left\{ \frac{\boldsymbol{\sigma}\cdot\boldsymbol{q}\,\boldsymbol{\sigma}\cdot\boldsymbol{k}\times\hat{\boldsymbol{\varepsilon}}}{2Mk} \left(\frac{1+\kappa_p+\kappa_n}{2} + \tau_3\, \frac{1+\kappa_p-\kappa_n}{2} \right) \right.\right.\right.$$

$$- \frac{\mathrm{i}\boldsymbol{\sigma}\cdot\boldsymbol{q}\,\boldsymbol{p}_f\cdot\hat{\boldsymbol{\varepsilon}}}{Mk}\cdot\frac{1+\tau_3}{2} \qquad\qquad (3.51)$$

$$\left.\left.\left. + \mathrm{i}\boldsymbol{\sigma}\cdot\hat{\boldsymbol{\varepsilon}}\left[\frac{1+\tau_3}{2} + \frac{k}{2M}\left(\frac{\kappa_p+\kappa_n}{2} + \tau_3\,\frac{\kappa_p-\kappa_n}{2} \right) \right] \right\} \tau_\alpha \chi_t^i \right| i \right\rangle .$$

For PV coupling the result is obtained by changing the coefficient of the last term:

$$\frac{1+\tau_3}{2} + \frac{k}{2M}\left(\frac{\kappa_p+\kappa_n}{2} + \tau_3\,\frac{\kappa_p-\kappa_n}{2} \right) \;\rightarrow\; \frac{E_\pi}{2M}\,\frac{1+\tau_3}{2} . \qquad (3.52)$$

The amplitudes are again found to consist of three terms. The first two terms are, as before, due to the particle component of the time-ordered diagrams (here the first graph in Fig. 3.6d), the third term due to the antiparticle graph. The essential differences are:

1) The second term survives if the final nucleon is a proton. It describes the effect of the interaction of the photon with the final nucleon. Here the second term does not vanish in the barycentric system.
2) The s-wave term is large in PS-coupling if the final nucleon is a proton (it is of order $1+k/2M$), whereas it is small (of order $k/2M$) if the final nucleon is a neutron. For PV coupling the s-wave term is small as before (of order $E_q/2M$).

The Complete Born Amplitude. From (3.43), (3.46) and (3.50) the total PS Born amplitude can be assembled. The result can be brought into the form (3.7):

$$T_{\mathrm{fi}} = \sum_{j=1}^{4} \bar{u}_f\, A_j(s,t)\, M_j u_i \quad . \qquad (3.53)$$

Using as the invariant operators M_i the set (3.10), the following expressions for the invariant amplitudes $A_j(s,t)$ are obtained:

$$A_1(s,t) = \tau_\alpha \left(\frac{1+\kappa_p+\kappa_n}{2} + \frac{1+\kappa_p-\kappa_n}{2}\,\tau_3 \right) \frac{ge}{s-M^2}$$

$$+ \left(\frac{1+\kappa_p+\kappa_n}{2} + \frac{1+\kappa_p-\kappa_n}{2}\,\tau_3 \right)\tau_\alpha \frac{ge}{u-M^2}$$

$$A_2(s,t) = -\tau_\alpha\,\frac{1+\tau_3}{2}\,\frac{ge}{(s-M^2)(t-M^2)} \qquad\qquad (3.54)$$

$$- \frac{1+\tau_3}{2}\,\tau_\alpha\,\frac{ge}{(u-M^2)(t-m^2)}$$

$$A_3(s,t) = \tau_\alpha \left(\frac{\kappa_p + \kappa_n}{2M} + \frac{\kappa_p - \kappa_n}{2M} \tau_3 \right) \tau_\alpha \frac{ige}{s - M^2}$$

$$A_4(s,t) = \left(\frac{\kappa_p + \kappa_n}{2M} + \frac{\kappa_p - \kappa_n}{2M} \tau_3 \right) \tau_\alpha \frac{ige}{s - M^2} .$$

Explicit forms for the amplitudes A_j corresponding to the set of operators (3.6) are, e.g., given in [3.12]. In (3.54) the invariant amplitudes are written as explicit functions of all three Mandelstam variables. However, since s, t, and u satisfy the relation

$$s + t + u = 2M^2 + m^2 \tag{3.55}$$

the invariant amplitudes can in principle be written as a function of only s and t. The PV Born amplitude differs from the PS version only by a small correction to A_1.

The amplitudes (3.54) are seen to consist of terms proportional to the following combinations of isospin operators

$$\tau_\alpha , \qquad \tau_\alpha \frac{1 \pm \tau_3}{2} , \qquad \frac{1 \pm \tau_3}{2} \tau_\alpha .$$

Using the relations

$$(1 \pm \tau_3) \tau_\alpha = \tau_\alpha \pm \frac{\{\tau_\alpha , \tau_3\}}{2} \mp \frac{[\tau_\alpha , \tau_3]}{2}$$

$$\tau_\alpha (1 \pm \tau_3) = \tau_\alpha \pm \frac{\{\tau_\alpha , \tau_3\}}{2} \pm \frac{[\tau_\alpha , \tau_3]}{2} \tag{3.56}$$

$$\frac{\{\tau_\alpha , \tau_3\}}{2} = \delta_{\alpha 3}$$

among the isospin operators, the decomposition (3.18)

$$A_j(s,t) = A_j^{(-)}(s,t) \frac{[\tau_\alpha , \tau_3]}{2} + A_j^{(+)}(s,t) \delta_{\alpha 3} + A_j^{(0)}(s,t) \tau_\alpha \tag{3.57}$$

of the invariant amplitudes in isospin space may be obtained.

Evaluating the isospin operators between the isospinors approxpriate for the different photopion channels the results in Table 3.2 are obtained.

Table 3.2. Isospin coefficients

	$\gamma p \to n\pi^+$	$\gamma n \to p\pi^-$	$\gamma p \to p\pi^0$	$\gamma n \to n\pi^0$
$\chi_i^{f\dagger}[\tau_\alpha , \tau_3]\chi_i^i/2$	$\sqrt{2}$	$-\sqrt{2}$	0	0
$\chi_i^{f\dagger}\{\tau_\alpha , \tau_3\}\chi_i^i/2$	0	0	1	1
$\chi_i^{f\dagger}\tau_\alpha \chi_i^i$	$\sqrt{2}$	$\sqrt{2}$	1	-1

It thus follows that terms proportional to the nucleon isospin transition operator τ_α contribute to both charged and neutral pion production, whereas terms associated with the isospin commutator (anticommutator) contribute only to charged (neutral) pion production. The amplitudes for specific reactions can be expressed in terms of the isospin amplitudes as

$$A_j(\gamma p \to n\pi^+) = \sqrt{2}\left(A_j^{(0)} + A_j^{(-)}\right)$$

$$A_j(\gamma n \to p\pi^-) = \sqrt{2}\left(A_j^{(0)} - A_j^{(-)}\right) \tag{3.58}$$

$$A_j(\gamma p \to p\pi^0) = A^{(0)} + A^{(+)}$$

$$A_j(\gamma n \to n\pi^0) = -A^{(0)} + A^{(+)} \quad .$$

It is thus apparent that only three of the four charged amplitudes are independent and the forth (e.g. that for $\gamma n \to n\pi^0$) can be expressed in terms of the other three. The isospin structure of photopion amplitudes as discussed above applies not only to the Born amplitudes but to all amplitudes.

The complete non-relativistic form of the (PS) Born amplitude is obtained by combining expressions (3.44), (3.47) and (3.51):

$$T_{\rm fi}^{\rm B,PS} = \frac{ef}{m}\left\langle {\rm f}\left| A^{(-)}\chi_t^{\rm f\dagger}\frac{[\tau_\alpha,\tau_3]}{2}\chi_t^{\rm i}\right.\right.$$
$$\left.\left. + A^{(+)}\chi_t^{\rm f\dagger}\frac{\{\tau_\alpha,\tau_3\}}{2}\chi_t^{\rm i} + A^{(0)}\chi_t^{\rm f\dagger}\tau_\alpha\chi_t^{\rm i}\right|{\rm i}\right\rangle \tag{3.59}$$

where

$$A^{(-)} = -\,{\rm i}\,\boldsymbol{\sigma}\cdot\hat{\boldsymbol{\varepsilon}} - \frac{{\rm i}\,\boldsymbol{\sigma}\cdot(\boldsymbol{k}-\boldsymbol{q})\boldsymbol{q}\cdot\hat{\boldsymbol{\varepsilon}}}{E_\pi k - \boldsymbol{q}\cdot\boldsymbol{k}} + \frac{1+\kappa_{\rm p}-\kappa_{\rm n}}{2}\frac{\hat{\boldsymbol{\varepsilon}}\cdot\boldsymbol{q}\times\boldsymbol{k}}{2Mk}$$
$$-\frac{{\rm i}\,\boldsymbol{\sigma}\cdot\boldsymbol{q}(\boldsymbol{p}_{\rm f}+\boldsymbol{p}_{\rm i})\cdot\hat{\boldsymbol{\varepsilon}}}{2Mk}$$

$$A^{(+)} = \frac{1+\kappa_{\rm p}-\kappa_{\rm n}}{2}\,{\rm i}\,\frac{\boldsymbol{\sigma}\cdot\boldsymbol{k}\,\boldsymbol{q}\cdot\hat{\boldsymbol{\varepsilon}}-\boldsymbol{\sigma}\cdot\hat{\boldsymbol{\varepsilon}}\,\boldsymbol{q}\cdot\boldsymbol{k}}{2Mk}$$
$$+\frac{k}{2M}\left(\kappa_{\rm p}-\kappa_{\rm n}\right){\rm i}\,\boldsymbol{\sigma}\cdot\hat{\boldsymbol{\varepsilon}}+\frac{{\rm i}\,\boldsymbol{\sigma}\cdot\boldsymbol{q}(\boldsymbol{p}_{\rm f}-\boldsymbol{p}_{\rm i})\cdot\hat{\boldsymbol{\varepsilon}}}{2Mk} \tag{3.60}$$

$$A^{(0)} = \frac{1+\kappa_{\rm p}+\kappa_{\rm n}}{2}\,{\rm i}\,\frac{\boldsymbol{\sigma}\cdot\boldsymbol{k}\,\boldsymbol{q}\cdot\hat{\boldsymbol{\varepsilon}}-\boldsymbol{\sigma}\cdot\hat{\boldsymbol{\varepsilon}}\,\boldsymbol{q}\cdot\boldsymbol{k}}{2Mk}$$
$$+\frac{k}{2M}\left(\kappa_{\rm p}+\kappa_{\rm n}\right){\rm i}\,\boldsymbol{\sigma}\cdot\hat{\boldsymbol{\varepsilon}}+\frac{{\rm i}\,\boldsymbol{\sigma}\cdot\boldsymbol{q}(\boldsymbol{p}_{\rm f}-\boldsymbol{p}_{\rm i})\cdot\hat{\boldsymbol{\varepsilon}}}{2Mk} \quad .$$

This result is valid in any coordinate system. The PV Born amplitude is obtained by replacing the terms in $A^{(+)}$ and $A^{(0)}$ proportional to $\boldsymbol{\sigma}\cdot\hat{\boldsymbol{\varepsilon}}$ by $(E_\pi/2M)\boldsymbol{\sigma}\cdot\hat{\boldsymbol{\varepsilon}}$.

3.6 Application of Dispersion Theory to Pion Photoproduction

According to the Kroll–Ruderman theorem [3.13] that the Born approximation becomes exact in the soft pion limit, and, as can be seen from Figs. 3.1 and 3.4, that it provides good estimates for the photoproduction amplitude near threshold, at least for charged pion production, the Born terms are dominant even for some energy region above threshold, but clearly as the resonance region is approached the Born approximation becomes strongly deficient. The situation is even less favorable for neutral pion production where the Born amplitude is small overall (Fig. 3.2) and the cross section is dominated by the non-Born amplitude already at 2 MeV above threshold.

The Born amplitude accounts for the poles in the photo-pion amplitude; an extension of the Born approximation requires evaluating the contributions of the other singularities in the amplitude: the branch cuts associated with intermediate multi-particle states in the s-, t- and u-channels.

Two approaches have been used successfully to accomplish this. The first one is a field-theoretical model based on the use of an effective Lagrangian, which results in the explicit evaluation of a number of Feynman diagrams: besides the Born diagrams a number of additional diagrams are evaluated, each of which corresponding to or approximating one of the multi-particle intermediate states in the s-, t-, and u-channels. The coupling constants involved are empirically determined to fit experimental data. This implies that the underlying Hamiltonian is an effective one. This in turn means that the amplitudes obtained can be used only to first order, i.e., they cannot be iterated. This approach is described in more detail in Sect. 3.7.

In the alternate approach which will be discussed in this section, the amplitude of the non-Born sector is obtained by iterating the Born amplitude to all orders. This approach based on the use of dispersion relations was pioneered by *Chew* et al. [3.8].

Applied to pion photoproduction, the dispersion theoretical approach is based on two assumptions:

1) The 12 invariant amplitudes $A_i^{(0)}$, $A_i^{(1)}$, $A_i^{(3)}$, $i = 1, \ldots 4$ as defined by (3.8), (3.18) and (3.19) are analytic functions of the Mandelstam variables s, t and u, and

2) the amplitudes are even or odd under exchange of s and u (crossing). Crossing symmetry follows from the assumed C-invariance of the S-matrix.

The singularities of the amplitudes are found to lie on the real axis of each variable: poles for each single intermediate particle and branch cuts for multiparticle intermediate states.

These properties are reflected in the structure of the resulting fixed-t dispersion relations:

$$\text{Re } A_j^{(n)}(s,t,u) = \left[\frac{1}{s-M^2} \pm \frac{1}{u-M^2} \right] \Gamma(t,u)$$

$$+ \frac{P}{\pi} \int_{(M+m)^2}^{\infty} ds' \text{ Im } A_j^{(n)}(s',t,u)$$

$$\times \left[\frac{1}{s'-s} \pm \frac{1}{s'-u} \right] . \tag{3.61}$$

The integration runs along a series of superimposed branch cuts on the positive real axis, starting at $s = (M+m)^2$, the location which corresponds to the position of the lowest intermediate multi-particle state. The first term in (3.61) was obtained by integration around the single pole at $s = M^2$, which corresponds to an intermediate single particle of mass M. The function $\Gamma(t,u)$ corresponds to the residue of the pole associated with the particle. The symbol P denotes the principal value of the integral, M and m are the nucleon and pion mass, respectively.

Dispersion relations are set up in the s-plane (for fixed t), since the resulting integral along the s-axis contains more interesting structure than an integral along the t-axis.

Equation (3.61) is the basic relationship of dispersion theory for pion photoproduction. One such relation exists for each of the 12 amplitudes $A_i(s,t,u)$; $i = 1,\dots 4$; $n = 0, 1, 3$, which together describe the photoproduction process. In the form (3.61), the dispersion relations are not very convenient to use. If unitarity is to be employed and the imaginary parts of the amplitudes are to be expressed in terms of pion scattering phase shifts, individual multipoles have to be projected out of each amplitude. The result is a set of four coupled equations for each pion partial wave and each of the three isospin configurations (0,1,3), relating the real parts of the partial wave amplitudes E_{l+}, E_{l-}, M_{l+}, M_{l-} to the imaginary parts of all partial wave amplitudes of the same isospin configuration.

In place of s as the integration variable it is more convenient to use the total barycentric energy $W = \sqrt{s}$. With the definitions

$$M_l(W) = \frac{W}{k\, q^l \sqrt{(W+M)^2 - m^2}} \begin{bmatrix} E_{l+} \\ e_f E_{l-} \\ M_{l+} \\ e_f M_{l-} \end{bmatrix} , \qquad e_f = E_f + M ,$$

k, q, and E_f being, respectively, photon and pion momentum, and final nucleon energy, the resulting integral equations can formally be written as [3.10]

$$\text{Re } M_l(W) = M_l^B(W) + \frac{P}{\pi} \int_{M+m}^{\infty} dW' \frac{\text{Im } M_l(W')}{W'-M}$$

$$+ \frac{1}{\pi} \sum_{l'} \int_{M+m}^{\infty} dW' \, K_{ll'}(W,W') \text{ Im } M_{l'}(W') . \tag{3.62}$$

The first term on the right hand side of (3.62) represents the contributions due to the projected Born terms, the two integrals give the corrections to the Born approximation. The first integral links the real and imaginary parts of the same multipole (direct term), the second one represents cross channel contributions and of s-channel terms beyond the direct term.

In the energy range below 500 MeV photon energy it is found [3.10] that the sum over l' in the third term in (3.62) can be limited to $l' = 0$ and 1. As long as the Watson theorem can be applied (which is the case, with small exceptions, also up to approximately 500 MeV), the imaginary parts of the multipole amplitudes can be expressed in terms of the real parts and the pion scattering phase shifts δ_l^π:

$$\text{Im } M_l = \text{Re } M_l \tan \delta_l^{\gamma\pi} = \text{Re } M \tan \delta_l^\pi \quad . \tag{3.63}$$

If this relation is inserted into (3.62), one obtains a well defined set of integral equations which can in principle be solved without any additional inputs beyond the Born terms and phase shifts.

The first attempt at solving the dispersion equations was presented by *Chew, Goldberger, Low* and *Nambu* in their classic paper [3.8] in 1957. The dispersion equations (3.62) were evaluated in the static limit and corrections to the Born amplitudes were evaluated to the lowest two pion partial waves, i.e., only corrections to the electric dipole amplitude E_{0+}, the magnetic dipole amplitude M_{1+}, and the electric quadrupole amplitude E_{1+} were considered significant. The Born terms themselves, on the other hand, were kept to all orders. The corrected amplitudes can thus be written as

$$F_1 = F_1^B + \Delta E_{0+} + 3 \cos \theta \left(\Delta M_{1+} + \Delta E_{1+} \right)$$

$$F_2 = F_2^B + 2\Delta M_{1+} + \Delta M_{1-} \tag{3.64}$$

$$F_3 = F_3^B + 3 \left(\Delta E_{1+} - \Delta M_{1+} \right)$$

$$F_4 = F_4^B \quad .$$

All corrections to the (0)-isospin channel were considered negligible. For the (−)-channel, which together with the (0)-channel determines the production amplitude for charged pions, the corrections were found to be

$$\Delta E_{1+}^{(-)} = \frac{i}{3} \, ef \, \frac{F_Q}{3} \left(2 e^{i\delta_{13}} \sin \delta_{13} + e^{i\delta_{33}} \sin \delta_{33} \right)$$

$$\Delta E_{0+}^{(-)} = \frac{i}{3} \, ef \, F_S (2\delta_1 + \delta_3) \tag{3.65}$$

$$\Delta M_{1+}^{(-)} = -\frac{i}{9} \, ef \, F_M \left(2 e^{i\delta_{13}} \sin \delta_{13} + e^{i\delta_{33}} \sin \delta_{33} \right)$$

$$\Delta M_{1-}^{(-)} = \frac{2i}{9} \, ef \, F_M \left(2 e^{i\delta_{11}} \sin \delta_{11} + e^{i\delta_{31}} \sin \delta_{31} \right)$$

Corrections to (+)-channel amplitudes are identical to the (−)-channel corrections except for a factor of (−2) multiplying all corrections.

The only terms under the dispersion integrals found to contribute to the corrections are the Kroll–Ruderman term and the photoelectric term — F_S, F_M, and F_Q being, respectively the electric and magnetic dipole and the electric quadrupole components of the combination of those two terms. Two recoil corrections were introduced: a term due to the current of a recoiling nucleon:

$$\Delta F^{(-,0)} = \mp i \, e f \, \frac{\boldsymbol{\sigma} \cdot \boldsymbol{q} \, \boldsymbol{q} \cdot \hat{\boldsymbol{\varepsilon}}}{2 M E_\pi} \tag{3.66}$$

and a factor $(1 + E_\pi/M)^{-1}$ multiplying the Kroll–Ruderman and photoelectric terms.

Since unitarity was built in at each intermediate step, the final result obtained by Chew et al. automatically satisfies the important unitarity requirement. Considering the early date of the CGLN amplitude, photopion cross sections predicted by it are in remarkable agreement with experimental results, fully justifying the classic status of the amplitude.

The approach to solve the dispersion equations (3.62) used in [3.10] was much more elaborate than that of [3.8]. All multipoles up to $l \leq 3$ were evaluated. The solutions obtained were in very good agreement with available data, the errors in general not exceeding 10 %, usually being considerably less than that. The only serious disagreements found were with low energy π^0-data.

The results obtained in [3.10] are provided in the form of tables of the multipole amplitudes $M_{l\pm}^{(0,1,3)}$ and $E_{l\pm}^{(0,1,3)}$ as a function of the photon laboratory energy for $l = 0, 1, 2, 3$. Since the complete solution is given in terms of multipoles and l is restricted to $l \leq 3$, this implies not only that the corrections to the Born approximation are given only up to $l = 3$, but that the Born approximation itself is limited to $l \leq 3$.

This is an aspect in which the photoproduction amplitude given in [3.10] is inferior to the CGLN amplitude. The correction terms to the Born approximation can certainly be expected to be negligibly small for $l > 3$ in the energy range considered. Similarly, it can be assumed that the Born terms due to the seagull diagram and to diagrams involving the nucleon propagator are well accounted for by four partial waves, since the main contributions here come from s- and p-wave, other partial waves entering essentially only through relativistic corrections.

In the photoelectric diagram, on the other hand, partial waves with $l > 3$ are clearly important, even at relatively low energies. A partial wave expansion of the photoelectric term shows that at a photon energy of 300 MeV only about 50 % of the forward amplitude is accounted for if the number of partial waves is limited to $l \leq 3$. At 200 MeV photon energy the deficiency in the forward direction still amounts to about 20 %. Thus if the amplitude of [3.10] is to be used for pion emission angles of less than 50° a sufficient number of partial waves of the photoelectric term with $l > 3$ have to be added. A simpler approach is to remove the lowest 4 partial waves of the photoelectric term from the total photopion amplitude and to add in their place the complete photoelectric term as, e.g., contained in the CGLN amplitude.

3.7 The Effective Lagrangian Approach

From a formal point of view the dispersion-theoretical approach discussed in Sect. 3.6 is the most satisfying way of extending the photopion amplitude beyond the Born approximation. However, even though this approach is in principle applicable at all energies, it is increasingly difficult to apply above the Δ resonance region. Also, as pointed out earlier, the amplitudes obtained in this approach are expressed in the pion-nucleon c.m. frame, and transforming them into other frames is cumbersome and ambiguous. In addition, it is difficult to discuss non-local and off-shell effects, which have turned out to play a significant role in nuclear applications of the elementary photopion amplitude.

An alternate approach which avoids some of these limitations is the effective Lagrangian model. In this approach the Born amplitude which is described by nucleon and pion pole diagrams is supplemented by additional pole terms with nucleonic resonances and vector mesons as intermediate states, i.e. the cuts in the s-, t-, and u-channels which describe the non-Born sector of the photopion amplitude are modeled by resonances with empirically adjusted coupling constants connecting the resonances to the initial and final states. The result is a phenomenological amplitude which has, however, the advantage that it can be readily expressed in a form that is valid in any coordinate frame. This fact was first recognized and exploited by *Blomqvist* and *Laget* [3.3].

Because of the dominance of the Δ resonance a very efficient parametrization of the non-Born sector is possible. In fact, for many applications a single diagram, that of s-channel Δ formation turns out to be sufficient. If in addition diagrams describing u-channel formation and ω^0 propagation in the t-channel are included, a very accurate description of the photopion amplitude from threshold through the Δ resonance region is possible. Furthermore, this approach can readily be extended to higher energies by simply including additional diagrams. It has also been used successfully to describe two pion production [3.14].

The effective Lagrangian approach is implemented by constructing a suitable chiral-invariant Lagrangian and evaluating the corresponding amplitude in the tree appproximation. The construction of an appropriate Lagrangian for pion photoproduction starts with the chiral Lagrangian for pion nucleon scattering [3.15]

$$\mathcal{L}_{\pi\mathrm{NN}} = \mathrm{i}\,\frac{f}{m}\,\overline{N}\,\gamma_\mu\gamma_5\tau^\alpha\,\mathrm{N}\,\partial_\mu\pi^\alpha \tag{3.67}$$

and with the Lagrangians $\mathcal{L}_{\mathrm{NN}}$ and $\mathcal{L}_{\pi\pi}$ for the free nucleon and pion fields. Chiral invariance implies PV coupling between nucleons and pions. Coupling of the photon to pions and nucleons is introduced by the minimal substitution

$$\partial_\mu \rightarrow \partial_\mu \pm \mathrm{i}eA_\mu \tag{3.68}$$

in $\mathcal{L}_{\mathrm{NN}}$, $\mathcal{L}_{\pi\pi}$ and $\mathcal{L}_{\pi\mathrm{NN}}$ and the addition of a term which describes the interaction of the photon field with the nucleon magnetic moments. The Lagrangian thus obtained corresponds precisely to the Born approximation, i.e., it accounts for the pion and nucleon pole terms.

As indicated earlier, branch cuts in the amplitude are accounted for in a phenomenological way by the addition of resonance terms. In particular, the pion-nucleon continuum is modeled by s-channel (and possibly u-channel) Δ formation. This is a reasonable assumption since the main pion-nucleon rescattering mechanism does indeed proceed through the $J = T = 3/2$ channel. To account for Δ formation two terms, $\mathcal{L}_{\gamma N\Delta}$ and $\mathcal{L}_{\pi N\Delta}$ are added to the Lagrangian, representing a phenomenological $\gamma N\Delta$ interaction (which couples the photon directly to the Δ resonance) and a corresponding $\pi N\Delta$ term describing the decay of the Δ. The complete chiral Lagrangian is thus [3.12]

$$\mathcal{L} = \mathcal{L}_{\gamma NN} + \mathcal{L}_{\gamma \pi\pi} + \mathcal{L}_{\pi NN} + \mathcal{L}_{\gamma NN\pi} + \mathcal{L}_{\Delta\pi N} + \mathcal{L}_{\Delta\gamma N} \quad . \tag{3.69}$$

The minimal substitution (3.68) ensures that the electromagnetic potential A_μ is coupled to a conserved current and therefore that for the Born part gauge invariance is preserved. For the general $\Delta\gamma N$ vertex two independent gauge-invariant couplings exist, one contributing to the (dominant) magnetic dipole $M_{1+}^{3/2}$, the other to the (small) electric quadrupole $E_{1+}^{3/2}$.

In the effective Lagrangian model, consistency with the low energy theorems predicted by current algebra is guaranteed by basing the approach on the chiral Lagrangian (3.67). In fact, the effective Lagrangian approach was originally conceived as an alternate simplified and more transparent way to obtain the soft pion results found by current algebra [3.16]. Current algebra obtains physical predictions from chiral systems through current commutation relations and use of the PCAC hypothesis. It was shown [3.15] that the same results can be obtained by using the lowest order graphs generated by any chiral-invariant Lagrangian.

An important aspect of the effective Lagrangian approach is that in evaluating the transition amplitudes only tree diagrams with appropriate external lines are to be summed over [3.17], i.e., diagrams with no self-connected points, requiring no integration after 4-momentum conservation of each vertex taken into account. The tree approximation applied to the Lagrangian (3.69) leads to the usual Born terms and in addition to diagrams with Δ propagators.

Peccei [3.18] used the chiral Lagrangian approach to derive the photopion Born amplitudes and to study pion photoproduction at threshold. *Olsson* and *Osypowski* [3.12] later extended applications of the approach to the Δ resonance region. They used Olsson's multi-channel approach [3.19] to unitarize their amplitudes. With their results they achieved remarkable agreement with experimental data. However, their photopion amplitude was again obtained in the pion-nucleon c.m. frame. Thus in order to use this amplitude in a nuclear calculation a frame transformation would be required.

Subsequently *Blomqvist* and *Laget* [3.3] applied the effective Lagrangian approach to derive a photopion amplitude which was directly applicable to an arbitrary frame of reference. This was an important advance since it eliminated in nuclear calculations the need for frame transformations to the frame of the moving nucleon and the ambiguities associated with this procedure. Furthermore, the diagrammatic structure of the amplitude facilitates the discussion of nonlocality and off-shell effects, which have turned out to play a significant role in nuclear applications. Blomqvist and Laget evaluated individual Feynman diagrams, as Olsson and Osypowski had done, but they evaluated the vertices and propagators in a frame independent way. They obtained a non-relativistic reduction of the photopion operator, valid to order p^2/M^2. In fact only the vertex operators, nucleon and Δ wave functions and the numerators of the propagators were evaluated to order $(p/M)^2$, whereas the full relativistic form was kept for the denominators of the propagators.

The Born Terms. The Born diagrams were evaluated as outlined in Sect. 3.5. The result obtained can be conveniently written in the following form:

$$\begin{Bmatrix} t_B(\gamma\pi^\pm) \\ t_B(\gamma\pi^0) \end{Bmatrix} = \frac{eg}{2M} \begin{Bmatrix} \sqrt{2} \\ 1 \end{Bmatrix} \{ D_1 i\sigma \cdot \hat{\varepsilon} + D_2 i\sigma \cdot \hat{\varepsilon}\, q \cdot k$$
$$+ D_3 i\sigma \cdot k\, q \cdot \hat{\varepsilon} + D_4 i\sigma \cdot q\, q \cdot \hat{\varepsilon} + D_5 i\sigma \cdot q\, p_i \cdot \hat{\varepsilon}$$
$$+ D_6 i\sigma \cdot q\, p_f \cdot \hat{\varepsilon} + D_7 q \cdot k \times \hat{\varepsilon} \} \quad . \tag{3.70}$$

The coefficients D_i are given in Table 3.3. They are expressed in terms of the pion propagator

$$P = \frac{1}{(q-k)^2 - m^2} \quad , \tag{3.71}$$

the factors

$$A = \frac{1}{2E_a\left(p_a^0 - E_a\right)} \quad , \qquad B = \frac{1}{2E_b\left(p_b^0 - E_b\right)} \tag{3.72}$$

stemming respectively, from the particle components of the propagators of the nucleon and crossed nucleon diagrams, the factors

$$\alpha = \frac{2M^2}{E_a\left(E_a + p_a^0\right)} \quad , \qquad \beta = \frac{2M^2}{E_b\left(E_b + p_b^0\right)} \tag{3.73}$$

Table 3.3. Coefficients of Born amplitude (PV coupling)

	$\gamma + p \to \pi^+ + n$	$\gamma + n \to \pi^- + p$	$\gamma + p \to \pi^0 + p$	$\gamma + n \to \pi^0 + n$
D_1	$-1 + \frac{q_0}{2m}\alpha$	$1 + \frac{q_0}{2m}\beta$	$(\alpha + \beta)\frac{q_0}{2m}$	0
D_2	$-A\mu_p + B\mu_n$	$-A\mu_n + B\mu_p$	$(B - A)\mu_p$	$(B - A)\mu_n$
D_3	$-A\mu_p + B\mu_n + P$	$A\mu_n + B\mu_p + P$	$(A - B)\mu_p$	$(A - B)\mu_n$
D_4	$-P$	$-P$	0	0
D_5	$-A$	0	$-A$	0
D_6	0	$-B$	$-B$	0
D_7	$A\mu_p + B\mu_n$	$A\mu_n + B\mu_p$	$(A + B)\mu_p$	$(A + B)\mu_n$

stemming from the corresponding antiparticle components, the fourth component q^0 of the 4-momentum q of the propagating pion, and the magnetic moments $\mu_p = 1 + \kappa_p = 2.78$ and $\mu_n = \kappa_n = -1.91$ of the proton and neutron. The quantities p_a and p_b are four-momenta of the propagating particles in the nucleon and crossed nucleon diagrams

$$p_a = \left(p_a^0, \boldsymbol{p}_a\right) \quad , \qquad p_b = \left(p_b^0, \boldsymbol{p}_b\right) \tag{3.74}$$

and

$$E_a = \sqrt{p_a^2 + M^2} \quad , \qquad E_b = \sqrt{p_b^2 + M^2} \tag{3.75}$$

the associated on-shell energies.

Since in the non-relativistic reduction only terms up to order $(p/M)^2$ were retained, expression (3.70) for the Born operator contains only 7 terms. If the formalism were to be extended to energies at which relativistic corrections become significant more or possibly all of the types of terms in (3.9) would have to be retained. If on the other hand the non-relativistic reduction is extended to the propagator terms, resulting in

$$\alpha = \beta = 1 \quad , \qquad A = B = \frac{1}{k} \tag{3.76}$$

the non-relativistic form of the Born amplitude given by (3.59) and (3.60) is recovered.

The s-Channel Δ-Exchange Diagram. The matrix element for this diagram (the last diagram in Figs. 3.6a and b) is taken to be [3.3]

$$T_{\mathrm{fi}}^{\Delta} = -\bar{u}_f\, C_\mu\left(p_f, p_\Delta\right) \frac{P^{\mu\nu}}{Q^2 - M_\Delta^2 - iM_\Delta\Gamma}\, \Gamma_\nu\left(p_i, p_\Delta\right) u_{\mathrm{i}} \tag{3.77}$$

where $p_\Delta = \left(p_\Delta^0, \boldsymbol{p}_\Delta\right)$ is the four-momentum of the intermediate Δ, $Q = p_\Delta^{0\,2} - p_\Delta^2$ the total pion-nucleon energy in the barycentric system, and M_Δ and Γ, respectivley, the mass and width of the Δ.

The general Δ propagator is complicated and contains "non-pole" terms [3.12]. Blomqvist and Laget use a simplified form and adopt as the projector

$$P^{\mu\nu} = 2M \sum_\Lambda u^\mu\left(\boldsymbol{p}_\Delta, \Lambda\right) u^\nu\left(\boldsymbol{p}_\Delta, \Lambda\right) \tag{3.78}$$

defined in terms of on-shell Rarita–Schwinger spinors. Since the Born terms are gauge-invariant, the s-channel Δ diagram has to be separately gauge-invariant. There are two couplings for the electromagnetic vertex which are separately gauge-invariant,

$$\Gamma_\nu\left(p_i, p_\Delta\right) = -C_\gamma\left[i\, G_1\left(\varepsilon_\nu - \frac{\gamma \cdot \varepsilon\, k_\nu}{M + M_\Delta}\right)\gamma_5\right.$$
$$\left. + \frac{i\, G_2\left(p_i \cdot \varepsilon\right)k_\nu - \left(p_i \cdot k\right)\varepsilon_\nu}{\left(M + M_\Delta\right)^2}\,\gamma_5\right] \quad . \tag{3.79}$$

The first coupling leads to the dominant magnetic dipole amplitude M_{1+}. The second coupling provides a large contribution to the E_{1+} electric quadrupole. However, since its overall contribution is only of order $(p/M)^2$ it is dropped.

In the non-relativistic limit the matrix element for the s-channel Δ formation becomes

$$T_{\mathrm{fi}}^{\Delta} = -\frac{C_{\pi}C_{\gamma}G_3G_1}{Q^2 - M^2 + iM_{\Delta}\Gamma}$$
$$\left\langle \mathrm{f} \left| \frac{2}{3} Q \cdot K \times \hat{\varepsilon} + i\sigma \cdot [Q \times (K \times \varepsilon)] \right| \mathrm{i} \right\rangle \tag{3.80}$$

with

$$Q = -q + \frac{q^0}{M_{\Delta}} p_{\Delta} \quad , \qquad K = k - \frac{M_{\Delta} - M}{M} p_i \quad , \tag{3.81}$$

where $q = (q^0, q)$ is the pion 4-momentum and C_{γ} and C_{π} are isospin factors which yield

$$C_{\gamma}C_{\pi} = \begin{cases} \pm\frac{\sqrt{2}}{3} & \text{for } \pi^{\mp} \text{ production} \\ \frac{2}{3} & \text{for } \pi^0 \text{ production} \end{cases} \tag{3.82a}$$

Mass and width of the Δ resonance as well as the pion-Δ coupling constant G_3 were found by fitting pion-nucleon scattering data:

$$M_{\Delta} = 1231 \text{ MeV}$$

$$\Gamma = 109 \left(\frac{|q|}{|q_{\Delta}|}\right)^3 \frac{M_{\Delta}}{Q} \frac{1 + (R|q_{\Delta}|)^2}{1 + (R|q|)^2} \text{ MeV} \tag{3.82b}$$

$$G = \frac{2.13}{m} \sqrt{\frac{1 + (R|q_{\Delta}|)^2}{1 + (R|q|)^2}} \text{ MeV}^{-1}$$

$$R = 0.00552 \text{ MeV}^{-1} \quad .$$

The electromagnetic coupling constant G_1 was obtained as

$$G_1 = 0.282 \frac{M_{\Delta} + M}{m} \sqrt{\frac{4\pi}{137}} \tag{3.82c}$$

by fitting charged pion photoproduction data.

The t-Channel ω^0 Exchange Diagram. The contributions of the t-channel cuts to the total amplitude are much less significant than those of the s-channel. In the t-channel, cuts are parametrized by vector meson exchange graphs. The major contribution here comes from ω^0 exchange, which is modeled by a t-channel pole term with the coupling constant for the vertex deduced from the radiative width of the ω^0, and the coupling constant of the ω^0NN hadronic coupling empirically determined so as to provide a good fit to the neutral pion production data. The ω^0 exchange amplitude which in the non-relativistic limit to order $(p/M)^2$ turns out to be [3.20]

$$T_{p\pi^0}^{\omega^0} = T_{n\pi^0}^{\omega^0} = \frac{1}{m} \frac{g_{\omega_1} g_{\gamma\pi\omega}}{(q-k)^2 - m_\omega^2} \, (q-k) \cdot \hat{\varepsilon} \times k \tag{3.83}$$

contributes mainly to the non-resonant magnetic dipoles which are not well reproduced by the Born terms alone. It vanishes for charged pion production.

The t-channel ϱ meson exchange amplitude should in principle also be considered. Its structure is the same as that for the ω^0 exchange. However, its contribution is negligible since both coupling constants are very small.

Assessment of the Blomqvist–Laget Model. The basic Blomqvist–Laget (B–L) model consists of the PV Born amplitude (3.70) to which the s-channel exchange contribution (3.80) and possibly the t-channel ω^0 exchange term are added. The resulting amplitude clearly violates the unitarity condition. However, it provides a satisfactory description of the major multipoles and it describes fairly accurately charged pion production cross sections. The major effect of the ω^0 exchange amplitude is to improve the fit to low energy neutral pion production data.

The non-unitarity of the amplitude manifests itself mainly in a poor description of the real part of the resonant $M_{1+}^{3/2}$ multipole amplitude above the resonance energy, which leads to a very poor reproduction of neutral pion production cross sections in that energy region. Following Olsson's approach [3.19], Blomqvist and Laget attempted to restore unitarity by modifying the resonant amplitude. They modified the resonance width by a form factor and introduced a phase factor in the electromagnetic coupling constant of the Δ. With the unitarity corrections included, the $M_{1+}^{3/2}$ multipole was reproduced to a high degree of accuracy, and the predicted cross sections based on this amplitude, particularly those for neutral pion production, were in much better agreement with available data.

In principle $E_{1+}^{3/2}$, the other multipole coupling to the Δ resonance, should also have been unitarized. However, by dropping the second coupling in (3.79), the B–L model ignores the resonance contribution to the $E_{1+}^{3/2}$ multipole altogether, keeping only the background contribution. Considering this approximation it is not surprising that the $E_{1+}^{3/2}$ multipole is not well reproduced in the B–L model. Since this multipole contributes only to order $(p/M)^2$, its poor representation is consistent with the level of accuracy the model is trying to maintain.

Despite the fact that a unitarized version of the B–L amplitude existed from the beginning, it was the non-unitarized form which has been widely used in nuclear calculations [3.21]. There are probably a number of possible explanations for this fact. One major reason may have been that the renormalization procedure was carried out in the barycentric frame (in which the multipoles are defined), which would have required additional frame transformations to make the result applicable in nuclear calculations. Another reason may have been that for charged pion production below the resonance region, where most calculations were carried out, the unrenormalized B–L amplitude is fairly accurate, and that the data then available were not of sufficient quality to pinpoint any deficiencies in the elementary amplitudes.

Overall, the major attraction of the B–L operator is that it is simple, that its diagrammatic structure provides a physically transparent picture of the individual processes which contribute, that it requires no ambiguous frame tranformations and that off-shell extrapolations are possible by simply replacing particle energies by the fourth components of off-shell momenta.

Efforts to Upgrade the Effective Lagrangian Model. In recent years the B–L amplitude has undergone some close scrutiny. Experimental results in nuclear pion photoproduction are now of sufficient quality that effects which contribute to calculations on the 10 % level are becoming significant. Furthermore, both theoretical and experimental investigations have concentrated on reactions such as $^{14}N(\gamma, \pi^+)$ $^{14}C_{g.s.}$ which strongly enhance the importance of otherwise insignificant multipoles. In addition, nuclear wave functions which are carefully calibrated against other reactions have become much more reliable. This has lead to a careful re-evaluation of the elementary photopion operator.

A number of attempts have been made to restore unitarity [3.22–26]. In these efforts the electric quadrupole is as a rule included in the unitarization process. The $E_{1+}^{3/2}$ multipole, although small, is an interesting case, since here the resonance component is small compared to the background, leading to strong interference effects and to large renormalization phases [3.23].

The problem of the frame dependence of the unitarization procedure seems to have been solved [3.26]. It should thus no longer be an obstacle to the acceptance of the unitarized amplitudes in nuclear calculations. More recently, *Laget* [3.23] has solved this problem by including a longitudinal $E2$ component and applying additional phases to satisfy unitarity. This amplitude gives good agreement with experiment for both charged and neutral pion production and is very suitable for application in a DWIA formalism.

A careful re-examination of the effective Lagrangian model by *Wittman* and *Mukhopadhyay* [3.26] leads to a number of refinements of the model: the second gauge coupling at the electromagnetic vertex of the s-channel Δ exchange diagram was restored, both resonant multipoles were unitarized, the Δ exchange in the u-channel was added, and some corrections in the non-relativistic reduction of the Born terms were included. These modifications to the elementary amplitude lead to substantially more accurate predictions of nuclear photoproduction cross sections.

In an effort to assess the effect of the non-relativistic reduction as obtained by the B–L model, *Bennhold, Tiator* and *Wright* [3.27] used the full relativistic production amplitude (3.8), with the operators given by (3.10), the invariant amplitudes for the Born part given by (3.54), and the corresponding invariant amplitudes associated with the s-channel exchange diagram. They thus essentially evaluated the elementary amplitudes with all the terms given by (3.9). Some terms in the transition operator associated with relativistic corrections are highly nonlocal and since nonlocal effects have been found to play a significant role in some nuclear photopion reactions, the effect of relativistic corrections to the B–L operator is enhanced in nuclear applications and may become significant. Even at the low energies (≤ 200 MeV) at which the calculations were carried out, the effect of relativistic corrections was quite noticeable. It would therefore be highly desirable to extend these investigations to higher energies where relativistic corrections are bound to be much more prominent.

The B–L model has been criticized that by using the projector (3.78) the Δ is not treated dynamically (the off-shell matrix elements are replaced by on-shell ones). *Tanabe* and *Ohta* [3.25] addressed this problem by writing the pion production amplitude in terms of a two-channel ($\gamma N, \pi N$) formalism. The πN amplitude is determined by coupling the πN and $\pi \Delta$ channels using separable potentials. Their approach yields an off-shell amplitude that could be used in nuclear applications. In this approach unitarity is built in from the start.

Araki and *Afnan* [3.26] extended the approach of Tanabe and Ohta. They obtain a unitary off-shell multi-channel theory of pion photoproduction including into consideration not only the Δ but also the Roper resonance. Their theory is based on the quark meson model (QMM). Thus the resonances are treated in terms of their quark content rather than in terms of their usual representation as π-N resonances. Coupling constants and vertex functions are calculated from a gauge and chirally invariant Lagrangian which is formulated in the language of the QMM. Since the Lagrangian used here is not an effective Lagrangian, propagators and vertices as well as coupling constants have to be renormalized. This is done in such a way that unitarity is satisfied.

Limitations of the Effective Lagrangian Approach. The main constraints for using elementary amplitudes such as those provided by the effective Lagrangian model in nuclear calculations come from the observation that the resonant part of the pion production mechanism proceeding through Δ excitation may be considerably modified in the nuclear medium, through effects such as nucleon Fermi motion, Pauli blocking of the Δ-decay, and the Δ-nucleus interaction.

The approach using effective Lagrangian operators provides a simple and accurate description of photoproduction of a pion from an on-shell nucleon. When computed in a nucleus, the Born terms are not expected to be strongly affected since they are determined by singularities in the elementary amplitude and constrained by gauge invariance. Thus medium corrections to the Born terms (with the possible exception of the pion pole term) can be assumed to be small.

In order to treat medium modifications one has to go beyond the impulse approximation and the use of an elementary production amplitude for a single nucleon is in principle no longer applicable.

A prominent case in charged pion production where medium effects are suspected to be responsible for discrepancies between DWIA calculations and experiments is the reaction $^{14}N(\gamma, \pi^+)^{14}C_{g.s.}$ which has recently received extraordinary attention, both experimentally and theoretically. *Tiator* et al. [3.28] use a hybrid model for the elementary production process to study this reaction. Instead of describing the resonant part of the production process by the s-channel Δ exchange diagram as done in the B–L model, they evaluate it in an extended version of the Δ-hole approach. The non-resonant background is described as in the B–L model by the PV Born term. The background contributions to the resonant $M_{1+}^{3/2}$ multipole stemming from the u-channel and the pion pole diagrams are projected out and substracted since they are already included in the Δ-hole description.

The results obtained using this model do not compare better overall with experiments than the DWIA calculation using the unitarized and upgraded B–L amplitude of [3.26]. In fact, they are only marginally better at one energy (320 MeV), but much worse at two other energies (200 and 260 MeV).

Suzuki, Takaki and *Koch* [3.29] also studied π^+ production on ^{14}N. Except for the pion pole term they dispensed with the effective Lagrangian approach altogether. Like Tiator et al. they treat the resonant production mechanism within the framework of the Δ-hole model to account for the major medium effects on the elementary production operator. The non-resonant production amplitude (with the exception of the pion pole contribution) was modeled by experimental single nucleon multipole amplitudes. Since the experimental multipoles are given in the pion-nucleon c.m. frame, a frame transformation was necessary. The results obtained (at 320 MeV) were similar to those of [3.28], i.e., also not significantly different from the DWIA result of [3.26]. It is thus still too early to decide whether (at least for charged pion production) medium effects are significant enough to justify abandoning the impulse approximation and thus the use of the full elementary pion production operator, or whether a properly unitarized and refined elementary operator is adequate for all but one or two exceptional cases.

3.8 Elementary Photopion Production at Higher Energies

Although the remainder of this book will deal with the photopion production process on complex nuclei at sufficiently low energies so that no more than the first (P_{33}) nucleon resonance plays any role in its mechanism or its interpretation, it might be of some interest here to discuss the elementary pion photoproduction process at some greater length, especially as it appears at higher energies. Figure 3.7 (from [3.30]) shows a diagram of the first and a few of the higher nucleon resonances, and indicates qualitatively their relative strengths in the elementary process $\gamma p \rightarrow \pi^+ n$, plotted vs. the photon energy E_γ.

Compilations of the relevant experiments can be found in papers analyzing the experimental results (e.g. [3.31]), in conference proceedings (e.g. [3.32]), and in the periodic publications of the Particle Data Group in Reviews of Modern Physics. Data analysis has been carried out e.g., for the $\gamma p \rightarrow \pi^+ n$ reaction by [3.31], by *Metcalf* and *Walker* [3.33], *Noelle* [3.34], and *Arai* [3.35], and for the $\gamma p \rightarrow \pi^0 p$ process by [3.33], by *Schwela* and *Weizel* [3.36], *Noelle* et al. [3.37], *Pfeil* and *Schwela* [3.38], and by *Moorhouse* et al.[3.39]. The models used in such

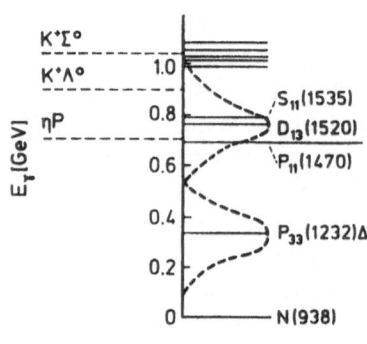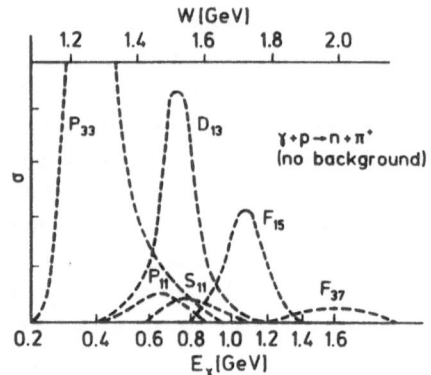

Fig. 3.7. Spectrum of nucleon resonances and their relative strength in the elementary process $\gamma p \rightarrow \pi^+ n$ [3.30]

analyses were phenomenological employing a decomposition into a Born term, Breit–Wigner resonances and additional "background" terms (*Walker* [3.33,40]), multipole analysis [3.38], partial waves [3.41], or they used fixed-t dispersion relations [3.31,42] and finite-energy sum rules [3.43]. More recently, the quark model [3.44], in particular the chiral bag model [3.45,46], has been applied to photopion production from a nucleon.

On a free proton target one can measure π^+ and π^0 photoproduction:

$$\gamma + p \rightarrow \pi^+ + n \quad , \tag{3.84a}$$

$$\gamma + p \rightarrow \pi^0 + p \quad , \tag{3.84b}$$

while measurement of the elementary amplitudes has also to include photoproduction on the neutron,

$$\gamma + n \rightarrow \pi^0 + n \quad , \tag{3.84c}$$

$$\gamma + n \rightarrow \pi^- + p \quad . \tag{3.84d}$$

The photon being an isospin mixture of $I = 0$ (isoscalar, s) and $I = 1$ (isovector, v) so that (3.84a) has an amplitude $\sim A^s + A^v$ while the amplitude of (3.84d) is $A^s - A^v$. To separate the A^s and A^v contributions, both of these reactions have to be measured. For obtaining the neutron target, a deuterium target can be employed:

$$\gamma + d \rightarrow \pi^0 + d \quad \text{(coherent)} , \tag{3.84e}$$

$$\gamma + d \rightarrow \pi^0 + n + p \quad , \tag{3.84f}$$

$$\gamma + d \rightarrow \pi^- + 2p \quad . \tag{3.84g}$$

Unfortunately, the analysis depends strongly on the structure of the deuteron (e.g., Fermi motion between p and n) and the final state interactions. Reaction (3.84e) was e.g. measured by *Von Holtey* et al. [3.47], and reaction (3.84g) by *Althoff* et al. [3.48].

In order to obtain complete information on all the amplitudes of the elementary reaction (3.84a–d), it is not sufficient to just measure cross sections and angular distributions, not even if single polarization measurements are added to this. It has been shown in a general fashion [3.49,50] that a total of nine measurements are needed to determine the amplitudes up to an overall phase, five of which must be double polarization measurements. These polarizations include polarized beams and polarized targets, and the measurement of recoil polarizations. The single polarization experiments possible comprise linearly polarized photon asymmetries (Σ), recoil baryon polarizations (P), and polarized target asymmetries (T). Of the twelve possible double polarization experiments, five must be chosen (in a particular combination) if all ambiguities except that of an overall phase are to be eliminated; this will be specified in more detail below. The definitive experiments which include polarization measurements are those by the Bonn group (see, e.g., [3.30]).

For these purposes, the Mandelstam variables are employed for describing the reactions (3.84a–d):

$$s = (\gamma + N_0)^2 \quad , \tag{3.85a}$$

$$t = (\gamma - \pi)^2 \quad , \tag{3.85b}$$

$$u = (\gamma - N)^2 \quad , \tag{3.85c}$$

$$v = \frac{s - u}{4m} \tag{3.85d}$$

where the particle symbols are written to stand for their four-momenta (N_0 = target nucleon, N = final nucleon); the analysis is always carried out in the CM-system where $s = w^2$, w being the CM total energy. One uses the parity-conserving t-channel helicity amplitudes F_i ($i = 1,\ldots,4$) which are related to the usual CGLN [3.8] invariant amplitudes A_i ($i = 1,\ldots,4$) as discussed earlier in this chapter, (3.5), by

$$F_1 = -A_1 + 2m\,A_4 \quad , \qquad F_2 = A_1 + t\,A_2 \quad ,$$
$$F_3 = 2m\,A_1 - t\,A_4 \quad , \qquad F_4 = A_3 \quad . \tag{3.86}$$

However, for further convenience one may introduce the s-channel helicity amplitudes S_1, S_2, N, and D where N is a no-flip amplitude, S_1 and S_2 single-flip amplitudes, and D a double-flip amplitude. The asymptotic crossing relation connects these amplitudes as follows:

$$\begin{pmatrix} F_1 \\ F_2 \\ F_3 \\ F_4 \end{pmatrix} = \frac{-4\sqrt{\pi}}{\sqrt{-t}} \begin{pmatrix} 2m & \sqrt{-t} & -\sqrt{-t} & 2m \\ 0 & \sqrt{-t} & \sqrt{-t} & 0 \\ t & 2m\sqrt{-t} & -2m\sqrt{-t} & t \\ 1 & 0 & 0 & -1 \end{pmatrix} \begin{pmatrix} S_1 \\ N \\ D \\ S_2 \end{pmatrix} \tag{3.87}$$

while the exact connection is provided by combining (3.86) with the formula of [3.10]. Alternately, one may work in terms of "transversity amplitudes"

$$b_1 = \tfrac{1}{2}\left[(S_1 + S_2) + i(N - D)\right], \quad b_3 = \tfrac{1}{2}\left[(S_1 - S_2) - i(N + D)\right],$$

$$(3.88)$$

$$b_2 = \tfrac{1}{2}\left[(S_1 + S_2) - i(N - D)\right], \quad b_4 = \tfrac{1}{2}\left[(S_1 - S_2) + i(N + D)\right].$$

To define the geometry, one adopts the usual Basel convention with the z axis being the beam direction and the y axis normal to the reaction plane, the z' axis being in the direction of the produced pion. If k is the incident photon momentum and q the outgoing pion momentum (using the c.m. system), then the axes are defined as

$$z = k/|k| \quad, \qquad y = (k \times q)/|k \times q| \quad, \qquad x = y \times z$$

$$(3.89)$$

$$z' = q/|q| \quad, \qquad y' = y \quad, \qquad x' = y \times z' \quad.$$

The sixteen observables in the elementary reactions are shown in Table 3.4 (from [3.50]), defined in terms of both helicity and transversity amplitudes.

Examples of polarization experiments as given in terms of these observables are:

Polarized Beam-Polarized Target

$$\frac{d\sigma}{dt} = \frac{d\sigma}{dt}\bigg|_{\text{unpol}} \Big\{1 - P_\text{T}\,\Sigma\,\cos 2\varphi + P_x\,(-P_\text{T}H\,\sin 2\varphi + P_\text{c}F)$$

$$- P_y\,(-T + P_\text{T}P\,\cos 2\varphi) - P_z\,(-P_\text{T}G\,\sin 2\varphi + P_\text{c}E)\Big\} \qquad (3.90a)$$

where (P_x, P_y, P_z) is the target polarization, P_T the transverse polarization of the beam at angle φ to the reaction plane, and P_c the degree of right circular polarization of the beam.

Target-Recoil

$$\varrho_\text{f}\,\frac{d\sigma}{dt} = \frac{d\sigma}{dt}\bigg|_{\text{unpol}} \Big\{1 + \sigma_y P + P_x\,(T_x\sigma_x + T_z\sigma_z)$$

$$+ P_y\,(T + \Sigma\sigma_y) - P_z\,(L_x\sigma_x - L_z\sigma_z)\Big\} \qquad (3.90b)$$

where

$$\varrho_\text{f} = \tfrac{1}{2}(I + \sigma \cdot P_\text{f}) \qquad (3.90c)$$

is the density matrix of the recoil nucleon, and P_f is its polarization.

The second-last column in Table 3.4 indicates the type of experiment required to determine the observables listed in the first column. These observables are shown to fall into several groups of four each: the first group (S) comprises the unpolarized cross section and the quantities Σ, T and P of single polarization experiments; after this follow three groups (denoted BT, BR and TR which stands for beam-target, target-recoil and beam-recoil) of double polarization experiments. It has been shown by *Barker* et al. [3.50] that all amplitudes can be determined unambiguously by measurement (up to an overall phase factor) if in

52

Table 3.4. Observables in single and double polarization experiments for the elementary reactions (3.84a–d).

Usual symbol	Helicity representation	Transversity representation	Experiment required [a]	Type
$d\sigma/dt$	$|N|^2+|S_1|^2+|S_2|^2+|D|^2$	$|b_1|^2+|b_2|^2+|b_3|^2+|b_4|^2$	$\{-;-;-\}$	
$\Sigma\, d\sigma/dt$	$2\mathrm{Re}(S_1^\dagger S_2 - ND^*)$	$|b_1|^2+|b_2|^2-|b_3|^2-|b_4|^2$	$\{L(\frac{1}{2}\pi,0);-;-\}$ $\{-;y;y\}$	
$T\, d\sigma/dt$	$2\mathrm{Im}(S_1 N^* - S_2 D^*)$	$|b_1|^2-|b_2|^2-|b_3|^2+|b_4|^2$	$\{-;y;-\}$ $\{L(\frac{1}{2}\pi,0);0;y\}$	S
$P\, d\sigma/dt$	$2\mathrm{Im}(S_2 N^* - S_1 D^*)$	$|b_1|^2-|b_2|^2+|b_3|^2-|b_4|^2$	$\{-;-;y\}$ $\{L(\frac{1}{2}\pi,0);y;-\}$	
$G\, d\sigma/dt$	$-2\mathrm{Im}(S_1 S_2^* + ND^*)$	$2\mathrm{Im}(b_1 b_3^* + b_2 b_4^*)$	$\{L(\pm\frac{1}{4}\pi);z;-\}$	
$H\, d\sigma/dt$	$-2\mathrm{Im}(S_1 D^* + S_2 N^*)$	$-2\mathrm{Re}(b_1 b_3^* - b_2 b_4^*)$	$\{L(\pm\frac{1}{4}\pi);x;-\}$	BT
$E\, d\sigma/dt$	$|S_2|^2-|S_1|^2-|D|^2+|N|^2$	$-2\mathrm{Re}(b_1 b_3^* + b_2 b_4^*)$	$\{c;z;-\}$	
$F\, d\sigma/dt$	$2\mathrm{Re}(S_2 D^* + S_1 N^*)$	$2\mathrm{Im}(b_1 b_3^* - b_2 b_4^*)$	$\{c;x;-\}$	
$O_x\, d\sigma/dt$	$-2\mathrm{Im}(S_2 D^* + S_1 N^*)$	$-2\mathrm{Re}(b_1 b_4^* - b_2 b_3^*)$	$\{L(\pm\frac{1}{4}\pi);-;x'\}$	
$O_z\, d\sigma/dt$	$-2\mathrm{Im}(S_2 S_1^* + ND^*)$	$-2\mathrm{Im}(b_1 b_4^* + b_2 b_3^*)$	$\{L(\pm\frac{1}{4}\pi);-;z'\}$	BR
$C_x\, d\sigma/dt$	$-2\mathrm{Re}(S_2 N^* + S_1 D^*)$	$2\mathrm{Im}(b_1 b_4^* - b_2 b_3^*)$	$\{c;-;x'\}$	
$C_z\, d\sigma/dt$	$|S_2|^2-|S_1|^2-|N|^2+|D|^2$	$-2\mathrm{Re}(b_1 b_4^* + b_2 b_3^*)$	$\{c;-;z'\}$	
$T_x\, d\sigma/dt$	$2\mathrm{Re}(S_1 S_2^* + ND^*)$	$2\mathrm{Re}(b_1 b_2^* - b_3 b_4^*)$	$\{-;x;x'\}$	
$T_z\, d\sigma/dt$	$2\mathrm{Re}(S_1 N^* - S_2 D^*)$	$2\mathrm{Im}(b_1 b_2^* - b_3 b_4^*)$	$\{-;x;z'\}$	TR
$L_x\, d\sigma/dt$	$2\mathrm{Re}(S_2 N^* - S_1 D^*)$	$2\mathrm{Im}(b_1 b_2^* + b_3 b_4^*)$	$\{-;z;x'\}$	
$L_z\, d\sigma/dt$	$|S_1|^2+|S_2|^2-|N|^2-|D|^2$	$2\mathrm{Re}(b_1 b_2^* + b_3 b_4^*)$	$\{-;z;z'\}$	

[a] Notation is $\{P_\gamma;\ P_T;\ P_R\}$ where:

P_γ = polarization of beam, $L(\theta)$ = beam linearly polarized at angle θ to scattering plane;

C = circularly polarized beam;

P_T = direction of target polarization;

P_R = component of recoil polarization measured.

In the case of the single polarization measurements we also give the equivalent double polarization measurement.

addition to the four experiments of type S, one carries out five double polarization experiments, of which no four of them should come from the same group. (No triple polarization experiments are needed for a full determination of the amplitudes). It is however possible to reduce the number of double polarization experiments to three (not all taken from the same group); in that case one may be left with a discrete eight fold ambiguity since each double polarization experiment leads to a trigonometric equation having a two-fold ambiguity in its solution. Reference [3.50] has shown that the mentioned conditions are necessary and sufficient for a complete determination of the amplitudes in the sense mentioned above. They also noted that double polarization measurements using a target polarized perpendicular to the reaction plane, or measuring recoil polarization perpendicular to the reaction plane, are equivalent to single polarization experiments as indicated in Table 3.4.

Extensive experiments on the elementary pion photoproduction reaction, including polarization experiments, have been carried out over the years at the Bonn 2.5 GeV synchrotron, the pions being produced by a bremsstrahlung beam incident on a liquid hydrogen target. Linearly polarized photons were produced [3.51] via coherent bremsstrahlung on a diamond target, based on the coherent bremsstrahlung method developed by *Überall* [3.52,53]. In the following, we show some results of the Bonn measurements.

Figure 3.8 presents the cross section data vs. photon energy of *Althoff* et al. [3.54] at c.m. pion angles $\theta_\pi^{c.m.} = 20°$ and $30°$, for photon energies K_γ from 500 to 1400 MeV. The solid curves correspond to the analysis by *Metcalf* and *Walker* [3.33]. It is interesting to note that the presence of the ηp channel threshold introduces a cusp into the cross section which has been analyzed by *Althoff* et al. [3.55], dashed line. Analyses using other models are also shown in [3.54].

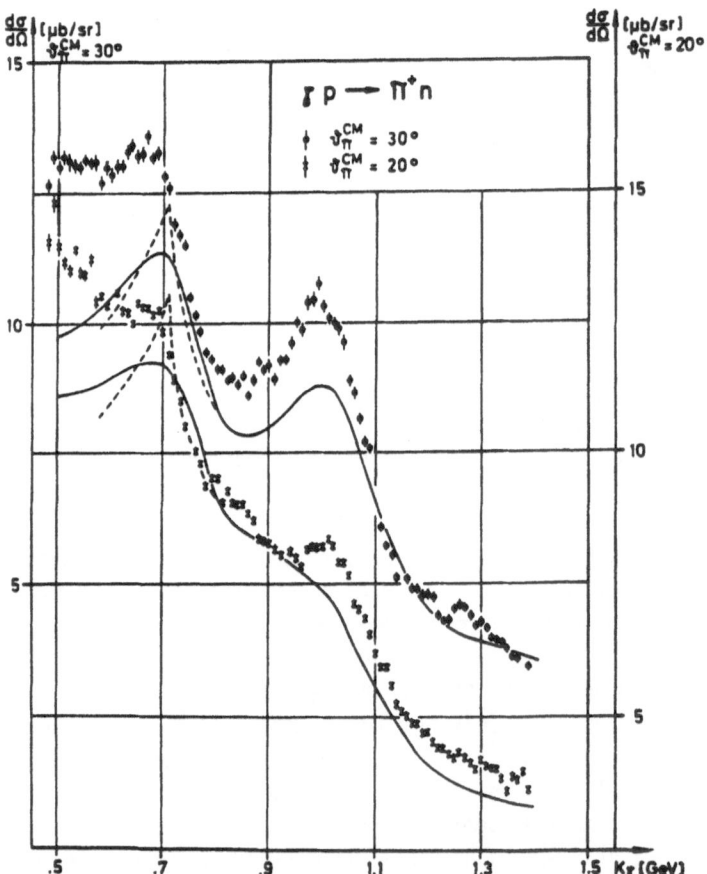

Fig. 3.8. Cross section at c.m.-pion angles 20° and 30° of the elementary process $\gamma p \rightarrow \pi^+ n$, as measured at Bonn [3.54], vs. photon energy K_γ. Solid curve: analysis of *Metcalf* and *Walker* [3.33]. Dashed curve: with η cusp effect included [3.55]

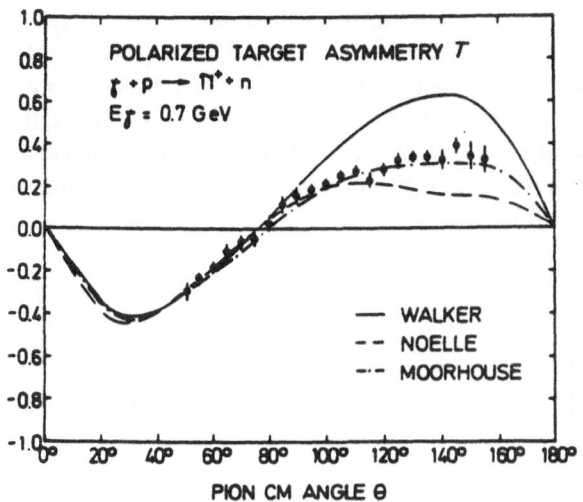

Fig. 3.9. Polarized target asymmetry $T\left(\theta_{\pi}^{c.m.}\right)$ of $\gamma p \rightarrow \pi^+ n$ at a photon energy $E_\gamma = 700$ MeV [3.58]. Analysis curves: see text

The target asymmetry T, with polarized protons, of the reaction $\gamma p \rightarrow \pi^+ n$ was measured by [3.56–58]. Figure 3.9 shows the angular dependence of T at a photon energy of 700 MeV [3.58]; the three curves represent the analysis of *Noelle* [3.61], *Moorhouse* et al. [3.39], and *Metcalf* and *Walker* [3.33], as indicated.

It has been possible to measure the reaction $\gamma n \rightarrow \pi^- p$ on a polarized neutron target (the neutron, of course, being imbedded in a deuteron) [3.60,62,63]. Since pure deuterium cannot be polarized dynamically, deuterated butanol doped with 5% normal water and 1% porphyrexide has been used. The deuterons are then polarized dynamically in a 2.5 T magnetic field at 0.5°K by an r.f. field, with 16% average deuteron polarization obtained; since 94% of the deuteron (spin $J = 1$) resides in the S-state, the neutron polarization is about that of the deuteron.

The target asymmetry T of $\gamma n \rightarrow \pi^- p$ measured in this way is shown in Fig. 3.10; the curves are given by the same analysis as in Fig. 3.9.

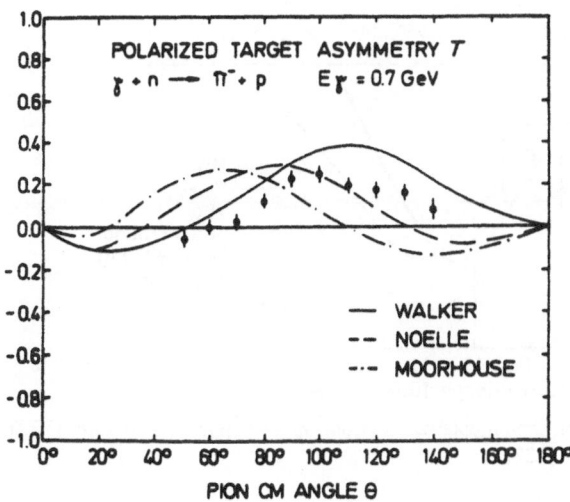

Fig. 3.10. Same as Fig. 3.9, for $\gamma n \rightarrow \pi^- p$ [3.63]

Neutral pion photoproduction on protons was measured in a series of experiments at Bonn [3.47,64-74]. In some cases, the target was polarized [3.73], or the polarization of the recoil proton was measured [3.71,72]. In Fig. 3.11, we show a combined set of Bonn data for the forward cross section, up to photon energies of 500 MeV [3.68], with an analysis curve by *Noelle* [3.61]. Figure 3.12 shows the recoil proton polarization in $\gamma p \rightarrow \pi^0 p$ [3.72] with curves corresponding to the Metcalf–Walker analysis [3.33].

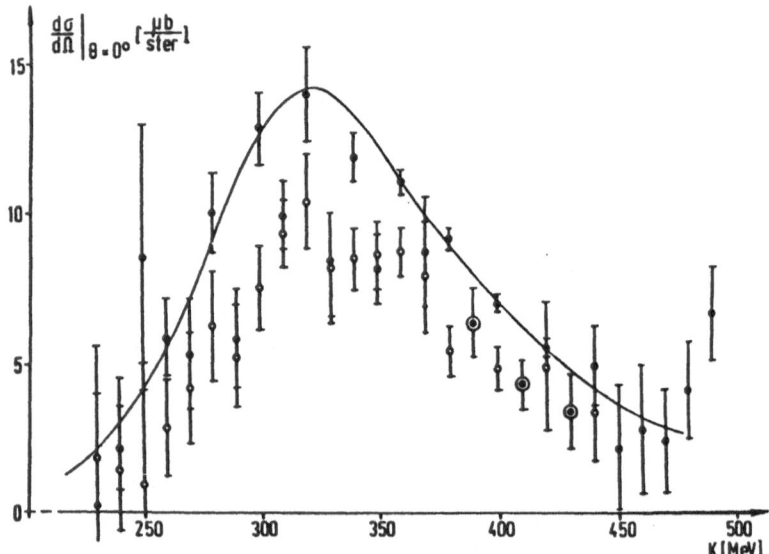

Fig. 3.11. Forward cross section of $\gamma p \rightarrow \pi^0 p$ up to 500 MeV photon energy, combining various Bonn data with an analysis by *Noelle* [3.68]

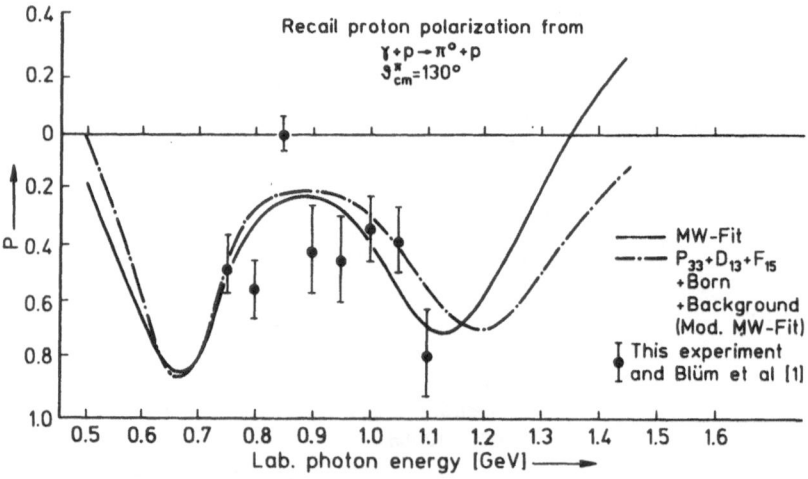

Fig. 3.12. Energy dependence of recoil proton polarization from $\gamma p \rightarrow \pi^0 p$ at $\theta_\pi^{c.m.} = 130°$ [3.72]. Comparison with Metcalf–Walker fits [3.33]

While pion photoproduction with real photons always corresponds to a vanishing photon mass $m_\gamma = 0$, electroproduction experiment such as

$$e + p \rightarrow e + n + \pi^+ \quad , \tag{3.91a}$$

$$e + p \rightarrow e + p + \pi^0 \quad , \tag{3.91b}$$

can be interpreted in terms of photoproduction experiments by virtual photons. For the latter, the "photon mass" is not zero, but corresponds to the magnitude of the 4-momentum transfer, $m_\gamma^2 = q^2$; moreover, three states of photon helicity ($\lambda = \pm 1$ *and* $\lambda = 0$) are possible, the latter corresponding to the Coulomb interaction of longitudinal photons (Fig. 3.13). These features add a new dimension to pion photoproduction.

Experiments on reactions (3.91a) [3.75–77] and (3.91b) [3.78,79] have also been carried out at Bonn. Some of these have been analyzed by the quark model. An extensive treatment of pion electroproduction is due to *Amaldi* et al. [3.80].

For a very complete polarization description of pion photoproduction, we refer to *Schmidt* and *Schwiderski* [3.81]. Finally, we might mention that in the biannual "Review of Particle Properties" under Baryon listings, one always finds a minireview of the latest status of pion photoproduction (in Sect. IV); see, e.g., [3.82].

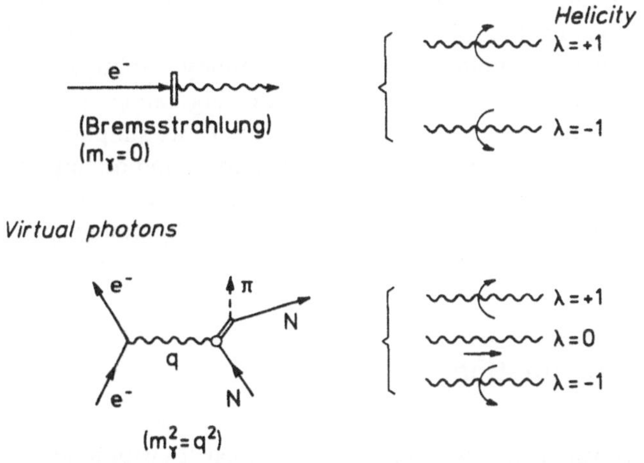

Fig. 3.13. Schematics of photoproduction (real photons) and electroproduction (virtual photons)

4. Nuclear Transition Amplitude

The major problem in photopion nuclear physics is the construction of the nuclear transition amplitude using the elementary photopion production amplitude on free nucleons. We use the condition that the elementary amplitude should yield the single nucleon cross section to a great degree of accuracy but this only tests the correctness of the on-shell behavior of the amplitude. The use of the elementary amplitude for the nuclear problem involves the extrapolation to the off-shell region and herein lies the greatest ambiguity. Besides, one has to take into account the initial and final state interactions of the probes with the nucleus. The interaction of the incident photon with the target nucleus is rather weak and hence can be neglected but the interaction of the outgoing pion with the residual nucleus is strong and should be included. Also the elementary amplitude treated as an operator for the nuclear problem may undergo modification in the nuclear medium and the multiple scattering effects in the nuclear medium may necessitate certain corrections for the simple impulse approximation that is used for the construction of the nuclear amplitude from the elementary amplitude. The nuclear transition amplitude can be constructed either in configuration space or in momentum space. The former has been used widely due to its simplicity in numerical calculation but the latter has a wider flexibility for incorporating the non-local effects. In this chapter, we shall discuss the construction of the nuclear transition amplitude using the impulse approximation both in configuration space and momentum space.

4.1 The Impulse Approximation

In this section, we shall consider how to construct the nuclear transition amplitude, given the single nucleon amplitude. This is done usually by assuming the impulse approximation and writing the nuclear transition operator T as a sum of free nucleon on-shell amplitudes t_i, where i denotes the nucleon index.

In the Fock space, the nuclear transition amplitude can be written in terms of annihilation and creation operators and one-body matrix elements:

$$T = \sum_i t_i = \sum_{\alpha,\alpha'} \langle \alpha'|t|\alpha \rangle \, a_{\alpha'}^+ \, a_\alpha \qquad (4.1)$$

where

$\alpha(\alpha')$: a set of quantum numbers $\{n, l, j, m, 1/2, m_\tau\}$ describing the bound nucleon,

$a_{\alpha'}^+$: particle creation operator above the Fermi energy ε_F, $\varepsilon > \varepsilon_F$, and

a_α : particle annihilation operator below the Fermi energy ε_F (or hole creation operator), $\varepsilon < \varepsilon_F$.

The impulse approximation involves the neglect of off-shell effects, multiple scattering in the nuclear medium, binding effects and the modifications of the free nucleon operator in the nuclear medium. At the first instance, we shall assume these effects to be small and later on incorporate certain corrections to take them into account.

In the impulse approximation, the nuclear photopion production amplitude can be written in terms of single nucleon photoproduction matrix elements:

$$\langle J_f M_f T_f N_f ; \pi |T| J_i M_i T_i N_i ; \gamma \rangle$$

$$= \sum_{\alpha,\alpha'} \langle J_f M_f T_f N_f | a_{\alpha'}^+ a_\alpha | J_i M_i T_i N_i \rangle \langle \alpha'\pi|t|\alpha\gamma\rangle \quad . \tag{4.2}$$

In equation (4.2), there is a clear separation of the nuclear structure part from the reaction mechanism of photoproduction of pions from single nucleons. The operator $a_{\alpha'}^+ a_\alpha$ can be expanded into irreducible tensor operators [4.1] in angular momentum–isospin space:

$$a_{\alpha'}^+ a_\alpha = \sum_{J,T} C(j' j J, m' - m M) C(\tfrac{1}{2} \tfrac{1}{2} T, m_{\tau'} - m_\tau M)$$

$$\times (-1)^{j-m}(-1)^{1/2-m_\tau} \left[a_{\alpha'}^+ \times a_\alpha \right]_{J,T}^{M,N} \quad . \tag{4.3}$$

The phase factors are due to the hole creation operator and are required to maintain the transformation properties of irreducible tensors. The nuclear structure part of the matrix element can now be written as

$$\langle J_f M_f T_f N_f | a_{\alpha'}^+ a_\alpha | J_i M_i T_i N_i \rangle = \sum_{J,T} (-)^{j-m} (-)^{1/2-m_\tau}$$

$$\times C(j' j J, m' - m M) C(\tfrac{1}{2} \tfrac{1}{2} T, m_{\tau'} - m_{\tau'} N)$$

$$\times C(J_i J J_f, M_i M M_f) C(T_i T T_f, N_i N N_f)$$

$$\langle J_f T_f \| \left[a_{\alpha'}^+ \times a_\alpha \right]_{J,T} \| J_i T_i \rangle \quad . \tag{4.4}$$

The quantity $\langle J_f T_f \| \left[a_{\alpha'}^+ \times a_\alpha \right]_{J,T} \| J_i T_i \rangle$ is the reduced matrix element of the one-body transition density corresponding to the single particle transition from state α to α' and it contains all the nuclear structure information.

If we restrict the consideration to closed shell nucleus in the initial state and one-particle one-hole nucleus in the final state, then the reduced matrix elements of one-body transition densities are just the configuration mixing coefficients $X_{ph}^{J_f T_f}$ of the final state nuclear wave function. Otherwise they depend on the configuration mixing coefficients of both the initial and final nuclear states.

The single particle matrix element of the photopion operator can be evaluated in the configuration space using a plane wave for the incident photon and a plane wave or a distorted wave for the outgoing pion. The use of a plane wave for the photon is justifiable but for the pion, the distortion effects are larger since the pion is a strongly interacting particle.

4.2 Plane Wave Impulse Approximation (PWIA)

If one uses the plane wave impulse approximation for both photon and pion, a closed expression can be obtained for the single nucleon matrix element.

$$Q = \langle \alpha' | t \exp(i \kappa \cdot r) | \alpha \rangle \tag{4.5}$$

where $\kappa (= k - q)$ is the momentum tranfer to the nucleon. The single nucleon transition operator has a general structure

$$t = \sigma \cdot K + L = \sum_{n=0,1} \sigma_n \cdot K_n \quad , \tag{4.6}$$

where $K (= K_1)$ is the spin-dependent part and $L (= K_0)$ the spin-independent part of the single nucleon photopion amplitude. For the purpose of combining both the parts of the amplitude and writing it in a more elegant form, a unit operator σ_0 is defined and the Pauli operator σ is redefined as σ_i. Expanding $\exp(i k \cdot r)$ into partial waves by using Rayleigh's expansion

$$\exp(i \kappa \cdot r) = 4\pi \sum_{l,m} (i)^l j_l(\kappa r) Y_{lm}(\hat{\kappa}) Y_{lm}(\hat{r}) \quad , \tag{4.7}$$

a closed expression for the single nucleon matrix element can be obtained [2.26].

The earlier calculations [2.26–28, 4.2–8] of photopion production cross sections from nuclei have been made using PWIA and they, in general, yielded cross sections which were much larger than the experimental values. In order to obtain agreement, a surface production model was introduced invoking the hypothesis that the pions produced well inside the nucleus were reabsorbed, thereby suppressing the pion production inside a core of radius r_0. In Fig. 4.1, the PWIA cross sections [2.26] obtained using the volume and surface production models are depicted along with the experimental data of *Meyer* et al. [2.23] for the reaction $^{16}O(\gamma, \pi^+)\,^{16}N(2^-, 0^-, 3^-, 1^-; T = 1)$. The final nuclear state is a group of low-lying levels with $J^\pi = 2^-, 0^-, 3^-, 1^-$ and $T = 1$. In the figure, V denotes the results of the volume production model, and S of the surface production model. The nuclear wave functions used are the independent particle model (IPM), those obtained by *Gillet* and *Vinh Mau* [4.9] using the random phase approximation [RPA] and the Migdal wave functions [4.10].

It has to be stressed that the surface production model is purely a phenomenological model that was used in the absence of any detailed knowledge of the final state interaction of the produced pion with the residual nucleus. Now considerable

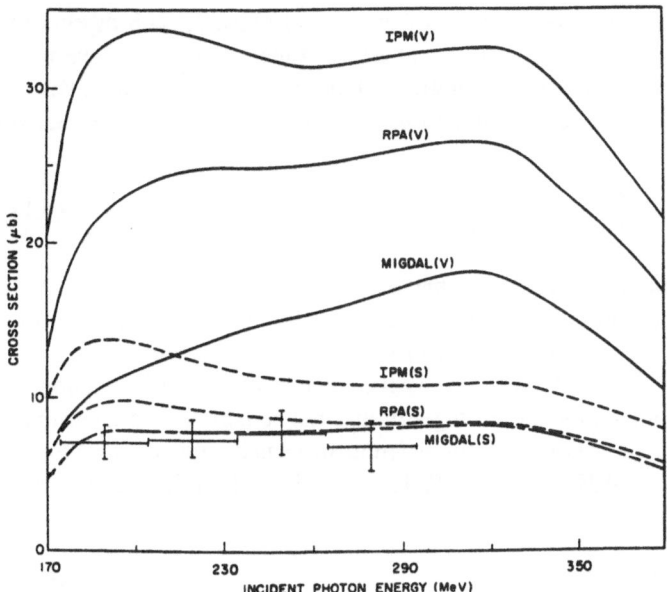

Fig. 4.1. Photoproduction of positive pions from ^{16}O leading to the low-lying states (J^{π} = $2^-, 0^-, 3^-, 1^-$; $T = 1$) of ^{16}N

progress has been made in the study of interactions of pions in nuclei by analyzing pionic atom data, elastic pion-nucleus scattering cross sections and pion absorption measurements. A reliable optical potential [2.37] for pion-nucleus interaction is now available for a wide range of energies up to 300 MeV and so it is possible to take into account the final state interaction of the pion with the nucleus in a more rigorous way.

4.3 Distorted Wave Impulse Approximation (DWIA)

In the distorted wave impulse approximation, distorted waves are used for the outgoing pion but a plane wave for the incident photon. The corresponding single nucleon matrix element is given by

$$Q = \langle \alpha' | \phi^*(q, r) \, t \, \exp(i k \cdot r) | \alpha \rangle \quad . \tag{4.8}$$

By making partial wave expansions of the outgoing pion wave function and the incident photon wave function, we obtain

$$\phi(q, r) = 4\pi \sum_{l_q, m_q} (i)^{l_q} \, g_{l_q}(\hat{r}) \, Y^*_{l_q m_q}(\hat{q}) \, Y_{l_q m_q}(\hat{r}) \tag{4.9}$$

$$\exp(i k \cdot r) = 4\pi \sum_{l_k \, m_k} (i)^{l_k} \, j_{l_k}(kr) \, Y^*_{l_k m_k}(\hat{k}) \, Y_{l_k m_k}(\hat{r}) \tag{4.10}$$

61

where $g_{l_q}(r)$ is the radial wave function of the partial wave l_q with asymptotic momentum q and it reduces to the spherical Bessel function $j_{l_q}(qr)$ if the pion-nucleus potential is switched off. The single nucleon transition operator is written in many equivalent forms by different authors in the evaluation of single nucleon matrix elements,

$$t = \sum_{n=0,1} \sigma_n \cdot K_n = \sum_i A_i O_i \quad . \tag{4.11}$$

The latter form is more convenient when the pion momentum in the single nucleon transition operator is replaced by a gradient operator [2.41] to take into account the variation of pion momentum inside the nuclear medium.

The evaluation of the single nucleon matrix element (4.8) involves a considerable amount of angular momentum algebra and the reduction has been made by different authors [2.38–41] in slightly different ways using one of the elementary amplitudes, viz., CGLN [2.15], BDW [2.30] and BL [2.45]. The transition operator for the nucleon

$$T = \phi^*(q, r)\, t \, \exp(i\mathbf{k} \cdot \mathbf{r}) \tag{4.12}$$

can be written as a scalar product of tensor operators of different ranks operating in nucleon space and tensors of the same rank constructed from the probes. Ultimately, the single nucleon matrix element is expressed in terms of single nucleon multipole transition densities.

Thus in the DWIA theory outlined above, the nuclear transition amplitude depends on the three ingredients: (i) the pion photoproduction operator on a single nucleon, (ii) the distorted pion wave functions and (iii) the appropriate nuclear transition densities. The free single nucleon photopion production amplitude has already been discussed in Chap. 3 and the other two ingredients viz., the distorted pion waves and the nuclear transition densities will be discussed in the subsequent chapters. In this chapter, we shall restrict our discussion to the ambiguities and uncertainties that are inherent in the application of free nucleon photopion production amplitudes to the nuclear problem.

Although CGLN and BDW amplitudes have been widely used [2.38,41], the BL amplitude is being increasingly used [2.39] in recent calculations. Despite the fact that the BL operator is well suited for nuclear applications, the procedure for using it in the nuclear problem is not unique.

One major uncertainty concerns the need to choose specific values for the zero components of the four-momenta of the final particles which appear in the denominators of the propagators. In the case of the pion propagator, one has e.g.,

$$(q - k)^2 - m_\pi^2 = q_0^2 - \left(q^2 + m_\pi^2\right) - 2q_0 k + 2\mathbf{q} \cdot \mathbf{k} \quad , \tag{4.13}$$

where q_0 is the energy of the final pion for which a value has to be chosen. A value of $q_0^2 = q^2 + m_\pi^2$ is the most commonly used one. However, other choices for q_0 may turn out to be preferable. Finding the optimum value for the zero components of the final particles is still an unsolved problem, which in any case can only be attacked in a proper fashion if it is done in conjunction with addressing the problem of medium modifications of the propagators.

Another problem to be addressed when using the elementary operator in a nuclear calculation concerns the fact that the momenta appearing in the transition operator are to be considered as operators acting on the appropriate wave functions. For calculations carried out in configuration space, this implies replacing nucleon and pion momenta by gradients. This, however, leads to the awkward situation of having to deal with gradients in the denominators of those terms in the elementary operator which contain propagators.

The standard solution to this problem is to apply the zero momentum approximation. This approximation in effect reduces the transition operator to a local or point operator. The consequences of this approximation depend strongly on the nucleus and the transition multipoles involved as will be shown below. The local approximation and the associated uncertainties can be avoided by carrying out all calculations in momentum space as outlined in Sect. 4.4 and this requires however the evaluation of the six-dimensional integrals leading to a very much larger computational effort [2.42].

Another source of uncertainty when using the BL operator in a nuclear calculation stems from the difference between the PS and PV versions of the operator. Whereas for charged pion production on free nucleons, the two coupling schemes lead to essentially equivalent results, for nuclear applications there may be significant differences in the results of a calculation that includes non-local effects. In PV pion-nucleon coupling [4.11], the usually dominant $\sigma \cdot \varepsilon$ term, as explained earlier, is primarily due to the Seagull diagram and is thus essentially a local term, whereas in PS coupling the Kroll–Ruderman (KR) term comes from the nucleon or crossed nucleon diagram which is associated with a propagator and is therefore sensitive to non-local effects.

The amount of uncertainty introduced into a non-local calculation by the difference between PS and PV coupling was investigated by *Toker* and *Tabakin* [2.46]. They found effects that were on average at the 20 % level but could grow significantly larger in the near-forward or backward directions.

Finally, an uncertainty exists concerning the Δ term in the BL operator. Although its presence is clearly necessary when the operator is applied to photoproduction on free nucleons, it has been argued repeatedly that the Δ terms should be dropped when the operator is used in nuclear calculations. In a growing number of cases where large discrepancies between theoretical predictions and data are encountered, it is found that significantly better agreement can be achieved if an elementary operator is used that does not contain the Δ term. Several of those cases will be discussed in Chap. 7.

The controversy concerning the Δ term stems partly from the fact that the phenomenological Δ diagram as contained in the BL operator includes the final state interactions of the pion with the active nucleon, which are also contained in the distorted pion waves used in the DWIA calculation. This double counting which appears to be a $1/A$ effect should clearly be corrected but it seems very unlikely that the correct way of doing so is to completely eliminate the Δ term from the elementary operator. As will be seen in Chap. 7, the cross sections calculated with and without the Δ term differ often very significantly, much more than would be expected from a $1/A$ effect.

Simply eliminating the Δ term appears to be wrong for other reasons also. It is well knwon that the Δ term plays a dominant role in π^0 production. Therefore eliminating it would very drastically reduce the nuclear (γ, π^0) cross section predictions and thus destroy the good agreement achieved between theory and experiment. Furthermore, in a recent paper by *Teng* et al. [4.12], the experimental cross sections obtained for the reaction $^{14}N(\gamma, \pi^+)\,^{14}C_{g.s.}$ at $E_\gamma = 320\,\text{MeV}$ were compared with the nonlocal calculations carried out with and without the Δ term. It was found that the cross section with the term included was already too low by a factor of three when compared with the data and that removing the Δ term would further reduce the cross section by another factor of four in the region of the delta resonance.

The question concerning the Δ operator is clearly a difficult one. In order to solve it, it is necessary to untangle the contribution of the elementary Δ (created by the incident photon exciting a nucleon) and the rescattering part (constituting the final state interaction of the elementary process), and furthermore and more importantly, to correctly treat the medium modifications of the Δ propagator and clearly separate them from the final state interactions of the pion with the nucleus which are described by the optical potential.

The DWIA calculations are most conveniently carried out in configuration space since then only the one-dimensional integrals have to be evaluated. This approach has been used in many calculations with considerable success. However, when for some nuclei, serious discrepancies between theory and data were found, the approximations made in the calculations had to be re-evaluated and it became clear that the most serious of these approximations was the pointlike treatment of the transition operator.

Toker and *Tabakin* [2.46] undertook an attempt to account for the nonlocality of the operator while retaining the simplicity of a configuration space calculation. Their approach was to replace the nonlocal diagrams by equivalent local diagrams by shifting the photon line to the other side of the propagators. All diagrams could thus be treated in configuration space. Nonlocalities were accounted for by replacing the pion waves $\psi_\pi(x)$ in the description of the pion pole term by effective pion waves

$$\tilde{\psi}_\pi(x) = \int d^3x' \, \Omega^\pi(x, x') \, \psi_\pi(x) \quad , \tag{4.14}$$

and the nucleon orbitals $\psi_{nl\mu}(x)$ in the description of the nucleon and Δ diagrams by the equivalent orbitals

$$\tilde{\psi}_{nl\mu}(x) = \int d^3x' \, \Omega^{N,N',\Delta}(x, x') \, \psi_{nl\mu}(x) \quad . \tag{4.15}$$

The smearing operator

$$\Omega(\boldsymbol{x} - \boldsymbol{x}') = \left\langle \boldsymbol{x} \left| \frac{1}{\beta^2 - m^2} \right| \boldsymbol{x}' \right\rangle \quad , \tag{4.16}$$

defined for each diagram as the Fourier transform of the associated propagator, has the effect of mixing pion partial waves (for the pion pole term) and of mixing nucleon orbitals (for the nucleon and Δ diagrams).

Toker and *Tabakin* [2.46] applied this formalism to π^- production on ^{12}C, ^{13}C and ^{14}N, all transitions leading to the ground states of the residual nuclei. They considered the contributions of the individual diagrams in the transition operator and how they are affected by nonlocalities. At low pion kinetic energies ($T_\pi < 50$ MeV), nonlocal effects were found to be very large on ^{14}N and also significant for ^{13}C, whereas they played a minor role for ^{12}C, the reason being that in ^{12}C the cross section is dominated by the local KR term. The nonlocal effects observed are found to be mostly due to the pion pole term.

The large gaps between theory and experiment which existed for both π^+ and π^- production on ^{13}C (and which in part prompted the investigation of nonlocalities) could not be closed by taking nonlocal effects into account. The reason for the large discrepancies between theory and data in these cases was later found to be mostly due to the use of insufficiently constrained nuclear transition densities and inaccurate data.

4.4 Formulation in Momentum Space

It is desirable to evaluate the single nucleon matrix element in momentum space since non-locality and off-shell effects can be handled better in this space. Following *Tiator* and *Wright* [2.42], the single nucleon matrix element in momentum space can be written as

$$\langle \alpha'\pi \,|\, t \,|\, \alpha\gamma \rangle = \int d^3p\, d^3q'\, \psi_{\alpha'}(\boldsymbol{p}')\, \phi_\pi^{-*}(\boldsymbol{q}', \boldsymbol{q})\, t(\boldsymbol{p}, \boldsymbol{p}', \boldsymbol{k}, \boldsymbol{q}')\, \psi_\alpha(\boldsymbol{p}) \,. \tag{4.17}$$

The wave function of the photon is a delta function in momentum space and the asymptotic momentum of the pion is denoted by $\boldsymbol{q}$. The momenta $\boldsymbol{p}$ and $\boldsymbol{q}'$ are the integration variables and $\boldsymbol{p}'$ is fixed by momentum conservation

$$\boldsymbol{k} + \boldsymbol{p} = \boldsymbol{q}' + \boldsymbol{p}' \quad . \tag{4.18}$$

The bound nucleon wave function $\psi_\alpha(\boldsymbol{p})$ is given by

$$\psi_\alpha(\boldsymbol{p}) = \sum_{m_l, m_s} C(l\tfrac{1}{2}j\,,\, m_l m_s m)\, \phi_{nlj}(p)\, Y_{lm_l}(\hat{p})\, \chi_{m_s}\, \tau_{m_\tau} \tag{4.19}$$

and a similar expansion can be made for $\psi_{\alpha'}(\boldsymbol{p}')$. Writing the one-body operator in the form

$$t = \sum_{n=0,1} \boldsymbol{\sigma}_n \cdot \boldsymbol{K}_n \tau^\mp = \sum_{n=0,1} \boldsymbol{\sigma}_n \cdot \boldsymbol{K}_n \tau^{-\beta} \beta/\sqrt{2} \tag{4.20}$$

where $\beta = +1$ for positive pion production and $\beta = -1$ for negative pion production, the single nucleon matrix element can be written in the following form after considerable amount of angular momentum algebra:

$$\langle \alpha' \pi \mid t \mid \alpha \pi \rangle = \sum_{m,m'} \sum_{L,n,J} I_M^{(a'a)LnJ} \begin{bmatrix} l' & l & L \\ \frac{1}{2} & \frac{1}{2} & n \\ j' & j & J \end{bmatrix}$$
$$\times C(j'\,j\,J,\, -m'\,m\,M)\, C(\tfrac{1}{2}\,1\tfrac{1}{2},\, m_\tau - \beta\, m_\tau')$$
$$\times \beta \sqrt{3}\, (-1)^{j'+m'-l'-n} \quad , \tag{4.21}$$

where $I_M^{(a'a)LnJ}$ denotes the six-dimensional integral involving the bound nucleon orbitals with quantum numbers $a = \{n, l, h\}$:

$$I_M^{(a'a)LnJ} = \int d^3p\, d^3q'\, \phi_\pi^{(-)*}(q', q)\, \phi_{n'l'j'}^*(p')\, \phi_{nlj}(p)$$
$$\times \left[\left[Y_{l'}(\hat{p}') \times Y_l(\hat{p}) \right]_L \times K_n \right]_J^M \quad . \tag{4.22}$$

The entire dynamics of the process is contained in the above six-dimensional integral. If we restrict to $1p$ shell nuclei and use harmonic oscillator wave functions with oscillator parameter b, the overlap of the single particle momentum wave functions is given by

$$\varrho_{a'a}(p', p) = \phi_{n'l'j'}^*(p')\, \phi_{nlj}(p)$$
$$= (8b^5/3\sqrt{\pi})\, pp'\, \exp\left[-(b^2/2)(p'^2 + p^2)\right] \quad . \tag{4.23}$$

The tensor operator $\left[\left[Y_{l'}(\hat{p}') \times Y_l(\hat{p}) \right]_L \times K_n \right]_L^M$ determines the strength of the various nuclear transitions.

For calculating the integral (4.22), we need a distorted pion wave function in momentum space. Starting from a standard solution $\phi_\pi^+(r, q)$ of an optical potential in coordinate space, we perform a Fourier tranform

$$\phi_\pi^+(q', q) = (2\pi)^{-3} \int d^3r\, \exp(-iq' \cdot r)\, \phi_\pi^+(r, q) \quad . \tag{4.24}$$

Expanding $\phi_\pi^+(r, q)$ and $\exp(-ip' \cdot r)$ into partial waves and integrating over angular variables, we obtain

$$\phi_\pi^+(q', q) = \frac{1}{2\pi^2} \sum_l (2l+1)\, P_l(q \cdot q') \int_0^\infty r^2 dr\, j_l(q', r)\, g_{l,q}(r) . \tag{4.25}$$

The above integral cannot be integrated numerically due to its behavior at infinity. So we need to follow a procedure outlined by *Tiator* and *Wright* [2.42].

Tiator and Wright have done the calculations without including the Coulomb effects, and all non-locel effects shown in their paper are obtained with the Coulomb potential neglected. Since their calculations correspond to pion energies of 50 MeV and above, neglect of the Coulomb potential may not matter much but at threshold energies, the neglect of the Coulomb potential will be very severe and hence will have to be included.

4.5 Beyond the Impulse Approximation

It has long been recognized that the DWIA is inadequate for pion scattering, which at intermediate energies is dominated by the formation and propagation of the Δ resonance. It was found that nuclear medium effects modify Δ propagation significantly [4.13]. Since neutral pion photoproduction also proceeds dominantly via Δ excitation, strong medium effects might be expected there, too.

If medium effects will turn out to play a significant role in charged pion production, they can be expected to be associated primarily with the propagator of the pion pole term, which has also turned out to be the source of the most significant non-local effects [2.46]. There are however cases of charged pion production where the Δ term provides the dominant contribution. Medium effects in these cases will be analogous to those encountered in neutral pion production.

Attempts have been made to account for medium effects, and thus go beyond the impulse approximation, in both charged and neutral pion photoproduction, although the approaches used were different. The approach chosen by *Dytman* and *Tabakin* [2.47] to describe the major medium modifications in charged pion production was to renormalize the nucleon spin operator based on ideas developed by *Chanfray* et al.[4.14] and *Mukhopadhyay* et al.[4.15] and applied to (ee′π) and (π, ππ) reactions by *Cohen* and *Eisenberg* [4.16]. All attempts to treat medium modifications in neutral pion production have so far been based on the isobar-doorway model [4.17] or its extension, the delta-hole model [4.18], both of which had previously been applied with success to pion-nucleus scattering.

The formalism developed by Dytman and Tabakin is based on the assumption that the major medium effects are due to virtual off-mass-shell pions exciting nucleon $p - h$ and $\Delta - h$ states while propagating. These effects were accounted for by renormalizing the spin operators of the KR term and the pion pole term.

The medium modifications to the transition operator as treated by Dytman and Tabakin lead to a large reduction of the cross sections in all cases considered, which were beneficial in some cases and not in others. In cases like the reaction $^{12}C(\gamma, \pi^+)^{12}B_{g.s.}$ at $T_\pi = 32$ MeV, which involves a nuclear transition that is dominated by the KR term, medium effects would be expected to be small. However, Dytman and Tabakin find with their formalism a reduction of the differential cross section by a factor of about two, drastically reducing the agreement with the data. For $^{14}N(\gamma, \pi^+)^{14}C_{g.s.}$ at $E_\gamma = 173$ and 200 MeV, where some medium effects might be expected, very strong reductions relative to the non-local results [2.46] are obtained. At the lower energy the agreement with the data is improved, whereas for the higher energy the cross section is reduced too much, and in addition the shape is wrong (see Fig. 4.2; non-local results: dot-dashed line, renormalized results: solid line). What is more serious is that the non-local results obtained with the momentum space code of *Tiator* and *Wright* [2.42] using the same nuclear information, give an excellent agreement with the data at both energies (as will be discussed in Chap. 7), suggesting that medium effects might be negligible or at least insignificant in these cases.

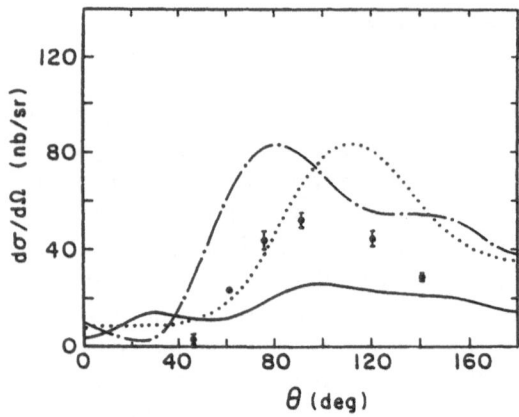

Fig. 4.2. Differential cross section for $^{14}\mathrm{N}(\gamma, \pi^+)\,^{14}\mathrm{C}_{\mathrm{g.s.}}$ at $E_\gamma = 200$ MeV. Local DWIA result [2.39] (*dotted curve*), non-local DWIA results [2.46] (*dot-dashed curve*), renormalized result [2.47] (*solid curve*) and experimental data from [4.22]

So far the only indication that the DWIA may be inadequate for charged pion production below pion kinetic energies of about 150 MeV comes from $E0$ transition. The large discrepancies between DWIA results and data and the unusual dominance of the Δ term in the transition amplitude make it appear likely that the DWIA is insufficient there. In the resonance region large gaps between theory and data remain to be resolved for a variety of reactions. Medium effects could possibly play a significant role in a number of these cases. So far, there is, however, little evidence that this is indeed the case.

For neutral pion production, on the other hand, the evidence that medium modifications are important is considerably stronger. In pion-nucleus scattering, effects due to nucleon and isobar binding, Pauli blocking of the Δ-decay by the nuclear medium and coupling of the propagating Δ to multi-hole (i.e. essentially absorption) channels were found [4.18]. Since neutral pion photoproduction is known to be dominated by Δ excitation, just as pion scattering is, Δ-related effects should be expected to be of considerable importance here also.

In order to treat medium induced Δ effects in coherent and incoherent neutral pion production, the dynamical description of the Δ-nucleus-system developed for pion-nucleus scattering (i.e. the isobar doorway model and more recently the Δ-hole formalism) has been applied. Since for neutral pion production the contribution of the Δ diagram strongly dominates, medium modifications affecting other diagrams are usually ignored, more specifically, the resonant part of the production amplitude is treated in the Δ-hole approach, the non-resonant part in DWIA.

Within the framework of the microscopic Δ-hole model the resonant amplitude for coherent π^0 production can be written as [4.19]

$$\langle q_i 0 |T_\Delta| k, \lambda_i 0 \rangle =$$

$$\left\langle q_i 0 \left| F_{\pi N\Delta}^\dagger \frac{1}{D(E - H_\Delta) - \delta W - W_\pi - V_{\mathrm{SP}}} F_{\gamma N\Delta} \right| k, \lambda_i 0 \right\rangle \qquad (4.26)$$

or alternatively as

$$\langle q_i 0 \,|\, T_\Delta \,|\, k, \lambda_i 0 \rangle =$$

$$\left\langle \psi_q^{(-)}; 0 \,\Big|\, F_{\pi N \Delta}^\dagger \, \frac{1}{D(E - H_\Delta) - \delta W - V_{SP}} \, F_{\gamma N \Delta} \,\Big|\, k, \lambda_i 0 \right\rangle \qquad (4.27)$$

where the effects of the pion scattering operator W_π (which describes intermediate pion propagation in the presence of the nuclear ground state) have been absorbed in the distorted wave $\langle \psi_q^{(-)}; 0|$ of the outgoing pion. The distorted pion wave function is calculated using the elastic pion-nucleus T-matrix, which in the Δ-hole approach is defined as [4.19]

$$T_{el}^{\pi\pi} = \left\langle q_i' 0 \,\Big|\, F_{\pi N \Delta}^\dagger \, \frac{1}{D(E - H_\Delta) - \delta W - W_\pi - V_{SP}} \, F_{\pi N \Delta} \,\Big|\, q_i 0 \right\rangle. \qquad (4.28)$$

H_Δ is the one-body Δ Hamiltonian. It is given by

$$H_\Delta = T_\Delta + V_\Delta + H_{A-1} \qquad (4.29)$$

where T_Δ is the kinetic energy operator of the Δ, V_Δ the binding potential for the Δ and H_{A-1} the Hamiltonian of the residual nucleus accounting for the hole energy. Pauli blocking of Δ decay in the nuclear medium is described by δW which reduces the free space decay width and shifts the resonance up in energy. The spreading potential V_{SP} is a phenomenological term which represents the coupling of the Δ propagator to intermediate channels with two or more nucleon holes, giving rise to pion absorption and higher-order multiple scattering effects. The parameters for the spreading potential are obtained from fits to pion-nucleus elastic scattering data. Pauli and binding corrections on the other hand are both evaluated microscopically in the space of Δ-hole configurations.

The (resonant) DWIA amplitude is obtained from (4.27) by setting $H_\Delta = T_\Delta$, $\delta W = 0$ and $V_{SP} = 0$ i.e. by replacing the many-body production operator based on the propagator of the Δ-hole state contained in (4.27) by the resonant part of the single-nucleon operator $t_{\gamma\pi}^\Delta$:

$$\langle q_i 0 \,|\, T_\Delta \,|\, k, \lambda_i 0 \rangle = \langle \psi_q^{(-)} 0 \,|\, t_{\gamma\pi}^\Delta \,|\, k, \lambda_i 0 \rangle \qquad (4.30)$$

with the pion wave function obtained from an optical potential instead of (4.28). In the Δ-hole approach described here, medium modifications are included both in the pion production operator and in the final state pion wavefunction. In fact, the resonant (i.e. the dominant) amplitudes for neutral pion production and pion scattering are here based on the same propagator, as seen from (4.26) and (4.28). This simply expresses the fact that in both cases it is assumed that the incident particle first excites a Δ in the nucleus, the so-called doorway state, and that the subsequent evolution of the system into its final state is determined by the Δ-nucleus interaction. A significant advantage of using the Δ-hole model is thus that the formalism developed to account for medium modifications in pion scattering can be directly applied to neutral pion production.

In [4.20] the Δ-hole approach is extended to treat incoherent photoproduction of neutral pions in which the nucleus is excited to a discrete final state.

Amplitudes are evaluated for transitions to particular ph states. Formally this is achieved by simply replacing in (4.27) the zero in the final state, indicating the nuclear ground state, by an appropriate particle-hole configuration.

The investigations on coherent π^0 production indicate that some of the medium corrections, e.g., those due to the spreading potential, indeed appear to be significant. One finds, however, that the contributions of the various medium corrections, while individually large, tend to cancel to a considerable extent, leaving only a relatively small net effect. For incoherent π^0 production, on the other hand, medium modifications of the transition operator are found to be considerably more involved than for coherent production leading in some cases to cross sections very significantly different from those obtained in DWIA [4.20].

The Δ-hole approach provides a valuable tool for analyzing medium effects. However, its application is by definition restricted to the resonant component of the transition operator and thus does not allow to account for medium effects to the "background" terms which are small only for neutral pion production near the resonance region. The Δ-hole approach thus provides at best only a partial solution of the problem of accounting for medium effects in pion photoproduction. How the nuclear medium affects the "background", i.e., the non-resonant terms of the elementary amplitude, in particular for charged pion production, is essentially an unanswered question. *Suzuki* et al.[4.21] have extended the study to the photoproduction of charged pions for which both resonant and non-resonant terms contribute. For the non-resonant terms, they have constructed a nuclear production operator based on the experimentally determined complex photopion multipoles for a free nucleon and an explicit treatment of the pion pole term. Nuclear photoproduction through the resonant channel was treated by the Δ-hole approach.

5. Pion-Nucleus Optical Potential and Distorted Pion Waves

The final state interaction of the outgoing pion with the nucleus is an important factor that affects the photopion cross sections from nuclei. The usual procedure is to replace the nucleus by an optical potential and consider the scattering of pions in this optical potential. The appropriate relativistic wave equation to be solved is the Klein–Gordon equation and the optical potential that is used should yield the pion-nucleus elastic cross section. The major problem is to choose the appropriate optical potential for the pion-nucleus interaction and solve the Klein–Gordon equation with this optical potential to obtain the distorted pion waves. In this chapter, we shall discuss the various optical potentials that are used and the solution of the Klein–Gordon equation.

5.1 Solution of the Klein–Gordon Equation

Distorted pion waves are obtained by solving the Klein–Gordon (K.G.) equation with an optical potential that is constrained to yield the pion-nucleus elastic cross section. The pion-nucleus elastic cross section can be studied either by solving the Lippmann–Schwinger equation in momentum space [5.1–3] or by solving the K.G. equation in coordinate space [1.1–3,5.4]. The first successful optical potential was that of *Kisslinger* [5.5] and it was non-local. *Fäldt* [5.6] obtained an equivalent local potential. *Krell* and *Barmo* (KB) [5.7] carried out a more phenomenological analysis of the scattering data by varying the dominant p-wave parameter in the optical potential. At low energies, the parameters of the optical potential were obtained by *Stricker* et al. (SMC) [2.37] from an analysis of the pionic atom data, nuclear absorption cross sections and pion-nucleus elastic cross sections.

To obtain the distorted pion waves, we need to solve the modified Klein–Gordon equation (omitting V_N^2 and $2V_c V_N$ terms)

$$\left(\nabla^2 + K^2\right)\psi(\boldsymbol{r}) = 2E\,V_N\,\psi(\boldsymbol{r}) \tag{5.1}$$

with

$$K^2 = \left(E - V_c\right)^2 - m_\pi^2 \quad . \tag{5.2}$$

The relativistic energy of the pion in the pion-nucleus barycentric system is denoted by E and the reduced mass of the pion by m_π. The Coulomb potential is V_c and the nuclear potential is V_N, for which let us assume the Kisslinger's gradient form [5.5]

$$2E\,V_N = q(r) - \nabla \cdot \alpha(r)\,\nabla \qquad (5.3a)$$
$$= q(r) - \nabla\,\alpha(r) \cdot \nabla - \alpha(r)\,\nabla^2 \qquad (5.3b)$$

where $q(r)$ denotes the local part and $\alpha(r)$, the non-local part of the nuclear potential. Substituting (5.3b) in (5.1), we obtain

$$\left[(1 + \alpha(r))\,\nabla^2 + \beta(r)\right]\psi(r) = -\nabla\,\alpha(r) \cdot \nabla\,\psi(r) \qquad (5.4)$$

where

$$\beta(r) = K^2 - q(r) \quad . \qquad (5.5)$$

If V_c, q and α are assumed to depend only on the radial coordinate, we can separate the radial wave equation from the angular parts by defining the wave function as a product of radial and angular functions

$$\psi(r) = R_l(r)\,Y_{lm}(\hat{r}) \quad . \qquad (5.6)$$

The radial wave equation so obtained is

$$(1 + \alpha(r))\left[\frac{1}{r^2}\frac{d}{dr}\left(r^2\frac{d}{dr}\right) - \frac{l(l+1)}{r^2} + \beta(r)\right]R_l(r)$$

$$= -\alpha'(r)\frac{d}{dr}\,R_l(r) \qquad (5.7)$$

which assumes a simpler form when the radial function $R_l(r)$ is transformed into $\tilde{u}_l(r)$

$$\tilde{u}_l(r) = \left(1 + \alpha(r)\right)^{1/2} r\,R_l(r) \quad . \qquad (5.8)$$

The resulting differential equation is

$$\left\{\frac{d^2}{dr^2} - \frac{l(l+1)}{r^2} - \frac{1}{1+\alpha(r)}\left[\frac{\alpha''}{2} + \frac{\alpha'}{r} - \frac{\alpha'^2}{4(1+\alpha)} - \beta(r)\right]\right\}\tilde{u}_l(r) = 0 \;. \;(5.9)$$

This equation can be solved by using the three-point Runge–Kutta method.

Beyond the range of nuclear potential, $q(r) = 0$ and $\alpha(r) = 0$ and (5.9) reduces to

$$\left\{\frac{d^2}{dr^2} - \frac{l(l+1)}{r^2} + K^2\right\}\tilde{u}_l(r) = 0 \quad . \qquad (5.10)$$

The general solution of this asymptotic wave equation is a linear combination of regular and irregular Coulomb functions

$$\tilde{u}_l(r) = F_l(\eta, Kr) + C_l\{G_l(\eta, Kr) + \mathrm{i}\,F_l(\eta, Kr)\} \quad . \qquad (5.11)$$

The charge parameter η is given by

$$\eta = \frac{EZZ'e^2}{m_\pi k} \quad , \tag{5.12}$$

where Z and Z' denote the charges of the incident pion and the target nucleus. The phase shifts δ_l are obtained in the usual way by matching the logarithmic derivatives of the inside and outside solutions at any point beyond the range of nuclear potential

$$C_l = e^{i\delta_l} \sin \delta_l \quad . \tag{5.13}$$

The differential cross section for elastic scattering is given by

$$\frac{d\sigma}{d\Omega} = |f_c(\theta) + f_N(\theta)|^2 \quad , \tag{5.14}$$

where $f_c(\theta)$ is the Rutherford scattering amplitude given by

$$f_c(\theta) = -\frac{\eta}{2K \sin^2(\theta/2)} \exp\left(-i\eta \ln \sin^2(\theta/2) + 2i\sigma_0\right) \tag{5.16}$$

and

$$f_N(\theta) = \frac{1}{K} \sum_{l=0}^{\infty} \exp(2i\sigma_l)(2l + 1) C_l P_l(\cos \theta) \quad . \tag{5.16}$$

The Coulomb phase shifts σ_l can be obtained by using the recurrence relation

$$\sigma_{l+1} = \sigma_l + \tan^{-1}\left(\frac{\eta}{l+1}\right) \quad , \tag{5.17}$$

starting from σ_0. For further details, the reader is referred to *Melkanoff* et al.[5.8].

Computer codes are available for solving the K.G. equation to obtain the distorted pion waves and also to calculate the pion-nucleus elastic cross section. Mention may be made of the code known as *Pirk* [5.4] developed at Pittsburgh and also another code developed by *Girija* and *Devanathan* (G.D.) [1.2,2.41] at Madras.

Instead of Kisslinger's gradient form, Fäldt's local form can also be used for the potential

$$2EV_N = q(r) + K^2 \alpha(r) + \frac{\nabla^2}{2} \alpha(r) \quad . \tag{5.18}$$

We now solve the corresponding radial equation

$$\frac{d^2 u_l(r)}{dr^2} + \left[\beta - K^2\alpha - \frac{1}{2} \nabla^2 \alpha - \frac{l(l+1)}{r^2}\right] u_l(r) = 0 \quad . \tag{5.19}$$

The wave functions u_l and $\tilde{u}_l$ obtained with the Kisslinger type and the Fäldt type of potentials are related through a factor $(1 + \alpha(r))^{1/2}$ and in turn to the radial wave function $g_l(r)$ defined by

$$u_l(r) = \frac{\tilde{u}_l(r)}{\left(1 + \alpha(r)\right)^{1/2}} = Kr\, g_l(r) \quad . \tag{5.20}$$

In the limit $V_N(r) = 0$ and $V_c(r) = 0$, $K \to q$ and

$$\tilde{u}_l(r) \to u_l(r) \to qr\, j_l(qr) \quad . \tag{5.21}$$

5.2 Pion-Nucleus Optical Potential

Normally one uses Kisslinger's gradient form (5.3a) for the pion-nucleus optical potential. The local part of the potential is denoted by $q(r)$ and the non-local part by $\alpha(r)$. *Krell* and *Barmo* (KB) [5.7] studied the elastic scattering of pions in the energy region $120 - 280$ MeV using the following form for the potential

$$q(r) = -4\pi \left[p_1 b_0 \varrho(r) - \varepsilon_\pi p_1 b_1 \delta\varrho + ip_2\, \mathrm{Im}\, B_0\, \varrho^2(r) \right] \tag{5.22}$$

$$\alpha(r) = \alpha_0(r) \left[1 - \tfrac{1}{3} \xi\, \alpha_0(r) \right] \tag{5.23}$$

$$\alpha_0(r) = -4\pi \left[p_1^{-1} c_0 \varrho(r) - \varepsilon_\pi p_1^{-1} c_1 \delta\varrho + ip_2^{-1}\, \mathrm{Im}\, C_0\, \varrho^2(r) \right] \tag{5.24}$$

$$\varrho = \varrho_p + \varrho_n \,, \quad \delta\varrho = \varrho_n(r) - \varrho_p(r) \,, \tag{5.25}$$

$$p_1 = 1 + \frac{m_\pi}{M} \,, \quad p_2 = 1 + \frac{m_\pi}{2M} \quad .$$

m_π denotes the rest mass of the pion and M the nucleon mass. The parameter ε_π assumes two values, $+1$ for π^+ scattering and -1 for π^- scattering, ξ is the Lorentz–Lorenz–Ericson–Ericson (LLEE) correction factor. The parameters b_0 and c_0 correspond to πN scattering lengths of s and p waves and the parameters b_1 and c_1 denote the corresponding πN scattering volumes. $\mathrm{Im}\, B_0$ and $\mathrm{Im}\, C_0$ are the parameters for the two-nucleon absorption. At the first instance, one can neglect the iso-vector terms and fix the values of the other parameters as follows

$$b_0 = -0.03\, m_\pi^{-1} \,, \quad \mathrm{Im}\, B_0 = 0.04\, m_\pi^{-4} \,, \quad \mathrm{Im}\, C_0 = 0.07\, m_\pi^{-6} \,, \quad \xi = 0 \quad .$$

These parameters are almost independent of the target nucleus with $\varrho_p = \varrho_n$ and hence describe well the scattering by ^{12}C and ^{16}O. The dominant parameter c_0 alone is found to be energy dependent and it can be varied as has been done by KB to obtain a good fit with experimental data. The proton and neutron density distribution ϱ_p and ϱ_n are normalized to Z and N respectively and for zero isospin nuclei,

$$\varrho_p(r) = \varrho_n(r) = N_c \left[1 + \frac{Wr^2}{a^2} \right] e^{-r^2/a^2} \,, \tag{5.26}$$

with the normalization constant

$$N_c = \frac{2Z}{\pi^{3/2} a^3 (2 + 3W)} \quad , \tag{5.27}$$

where $W = (A - 4)/6$ and a is the radius parameter. For heavier nucleus such as ^{40}Ca, it is better to assume a Woods–Saxon distribution given by

$$\varrho(r) = \frac{\varrho_0}{1 + \exp\left(\frac{r-c}{t}\right)} \quad , \tag{5.28}$$

with

$$\varrho_0 = \frac{3}{4\pi c^2 \left[1 + \frac{\pi^2 t^2}{c^2}\right]} \quad . \tag{5.29}$$

In the above equations c denotes the size parameter for nuclear matter density and t is the thickness parameter. The Coulomb potential V_c is calculated assuming the nucleus to correspond to a sphere of radius R_c with uniform charge density

$$R_c = R_{0c} A^{1/3} \quad , \tag{5.30}$$

where R_{0c} denotes the Coulomb radius constant.

$$V_c(r) = \frac{ZZ'e^2}{2R_0} \left[3 - \frac{r^2}{R_c^2}\right] \quad , \quad \text{for} \quad r \leq R_c \quad ,$$

$$= \frac{ZZ'e^2}{r} \quad , \qquad \text{for} \quad r \geq R_c \quad . \tag{5.31}$$

The parameters b_0, c_0, b_1 and c_1 can, in general, be obtained from $\pi - \text{N}$ phase shifts [5.9]

$$b_0 = \frac{1}{3}\left(\alpha_{11}^0 + 2\alpha_{31}^0\right) \tag{5.32}$$

$$c_0 = \frac{1}{3}\left(\alpha_{11}^1 + 2\alpha_{13}^1 + 2\alpha_{31}^1 + 4\alpha_{33}^1\right) \tag{5.33}$$

$$b_1 = \frac{1}{3}\left(\alpha_{31}^0 - \alpha_{11}^0\right) \tag{5.34}$$

$$c_1 = \frac{1}{3}\left(\alpha_{31}^1 + 2\alpha_{33}^1 - \alpha_{11}^1 - 2\alpha_{13}^1\right) \quad . \tag{5.35}$$

The quantity α in (5.32–35) depends upon the pion-nucleon scattering phase shifts δ_l.

$$\alpha_{2T,2J}^l = \exp(i\delta_l) \sin \delta_l / \mu_{\text{cm}}^{2l+1} \quad , \tag{5.36}$$

where μ_{cm} denotes the pion momentum in the centre of mass of the pion-nucleon system. The parameters are then Lorentz transformed to the pion-nucleus centre of mass by multiplying them by the factor $(\omega_\pi \omega_N)/(E_{\text{lab}} M)$ where ω_π and ω_N are the energies of the pion and nucleon respectively in the centre of mass of the pion-nucleon system and E_{lab} is the energy of the pion in the laboratory system.

It may be observed that the pion-nucleus centre of mass system approximately coincides with the laboratory system.

However, at low energies, the pion-nucleon optical potential described by (5.23–25) is found to be inadequate and extra terms of the form $\nabla^2 \varrho$ and $\nabla^2 \varrho^2$ were found to be necessary. *Stricker, McManus* and *Carr* (SMC) [2.37] have developed a potential which works quite well at low energies (below 50 MeV) by adding additional terms and fixing the optical model parameters by a compromise fit to data from pionic atoms, elastic scattering and absorption measurements.

$$q(r) = -4\pi \left[p_1 b(r) + p_2 B(r) + \frac{1}{2}\left(1 - p_1^{-1}\right) \nabla^2 c(r) \right.$$

$$\left. + \frac{1}{2}\left(1 - p_2^{-1}\right) \nabla^2 C(r) \right] , \tag{5.37}$$

$$\alpha(r) = 4\pi \frac{L(r)}{1 + \frac{4\pi}{3}\, \xi\, L(r)} , \tag{5.38}$$

with

$$b(r) = \bar{b}_0 \varrho(r) - \varepsilon_\pi b_1 \delta\varrho , \tag{5.39}$$

$$L(r) = p_1^{-1} c(r) + p_2^{-1} C(r) , \tag{5.40}$$

$$c(r) = c_0 \varrho(r) - \varepsilon_\pi c_1 \delta\varrho , \tag{5.41}$$

$$B(r) = B_0 \varrho^2(r) - \varepsilon_\pi B_1 \varrho\delta\varrho , \tag{5.42}$$

and

$$C(r) = C_0 \varrho^2(r) - \varepsilon_\pi C_1 \varrho\delta\varrho . \tag{5.43}$$

The parameters B and C indicate terms arising from true pion absorption, while parameters b and c denote terms arising from single nucleon scattering. Isoscalar and isovector terms are distinguished by the subscripts zero and one, respectively. ξ is the LLEE correction factor. The kinematic factors are defined by

$$p_1 = \frac{1 + \varepsilon}{1 + \varepsilon/A} , \quad p_2 = \frac{1 + \varepsilon/2}{1 + \varepsilon/2A} , \quad \varepsilon = E_\pi/M , \tag{5.44}$$

where E_π denotes the relativistic energy of the pion and M the nucleon mass.

Although the form of the potential is chosen from theoretical considerations, it has too many parameters to be determined from fits to the experimental data. The available data are, however, not sufficient to uniquely fix the values of all parameters in the potential as a function of pion energy. Only three or four parameters can be determined at a time by fits to data, the other parameters have to be taken from theoretical arguments, which at this point, are not reliable.

5.3 Pion-Nucleus Elastic Scattering

The cross sections for elastic scattering of π^+ and π^- by several nuclei have been calculated using the afore mentioned pion-nucleus optical potential and compared with the available experimental data [5.10–22]. Reasonable agreements have been obtained both at low and medium energies.

The effect of the $\nabla^2 \varrho(r)$ and $\nabla^2 \varrho^2(r)$ terms in the SMC potential on the cross sections is depicted in Fig.5.1 for 28.4 MeV and 38.7 MeV for $\pi^+ - {}^{12}$C scattering. It is seen that the inclusion of the above terms enhances the back angle scattering, thereby improving the agreement with the experimental data. Also the effect of these terms is to shift the minima. However these terms do not appreciably alter the cross sections at medium energies. An interesting feature of the pion is that the change in the charge can alter the Coulomb-nuclear interference. Fig. 5.2 depicts the π^+ and π^- cross section with ^{12}C at 50 MeV and it is possible to estimate the strength of the Coulomb-nuclear interference in the potential. The angular distribution obtained with the SMC potential for 120 MeV $\pi^- - {}^{12}$C elastic scattering is in excellent agreement with the data of *Binon* et al. [5.12] (see Fig. 5.3).

However, it has to be observed that the agreement of the calculated cross sections with the pion-nucleus elastic scattering data does not ensure the correctness of the pion-nucleus optical potential. It is true that the cross sections for elastic scattering of pions from nuclei are obtained from the solution of the

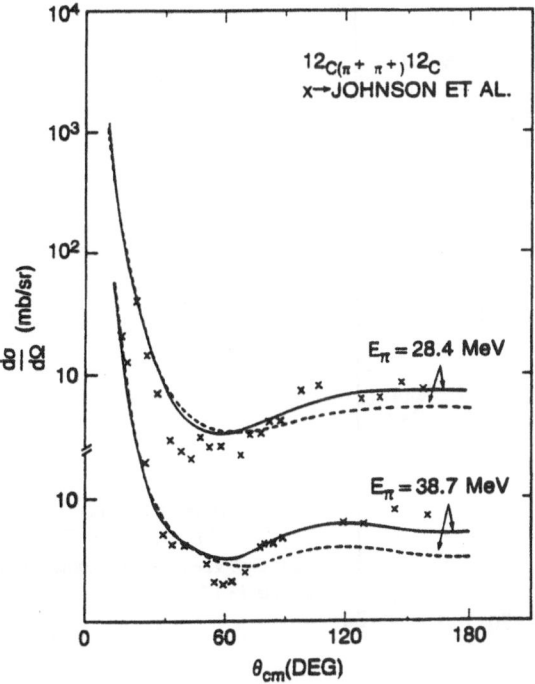

Fig. 5.1. Elastic scattering of π^+ by ^{12}C, compared with the data of Johnson et al. [5.21]. The continuous and dashed lines are those obtained respectively with and without $\nabla^2 \varrho(r)$ and $\nabla^2 \varrho^2(r)$ terms in the SMC potential [2.37]

77

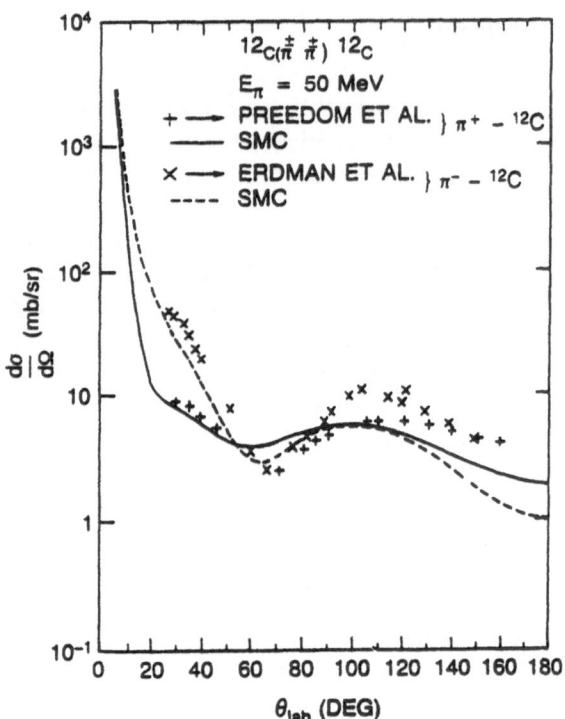

Fig. 5.2. Comparison of the elastic scattering of π^+ and π^- on ^{12}C for laboratory pion energy $E_\pi = 50$ MeV. The data for π^+ and π^- scattering are taken from [5.11, 21] respectively

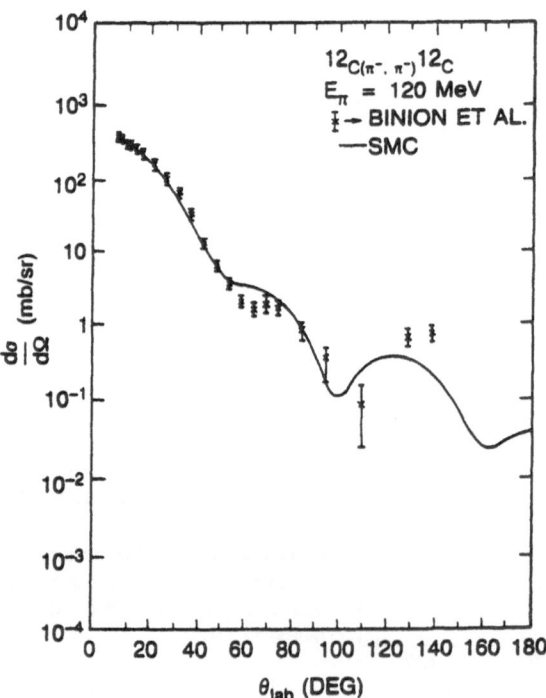

Fig. 5.3. Differential cross sections for $\pi^- - {}^{12}C$ scattering for $E_\pi = 120$ MeV compared with the experimental data [5.12]

Klein–Gordon equation using the optical potential. But the elastic cross section depends essentially on the asymptotic wave function of the pion. Since it is possible to generate the same phase shift from two different optical potentials, the elastic cross section cannot distinguish between two different phase equivalent optical potentials [5.23]. The inelastic pion scattering and photoproduction of pions will, in principle, distinguish between them since the DWIA formalism uses the complete wavefunction. But the problem becomes complicated due to uncertainties in medium modifications of the elementary transition operator and the nuclear wave functions. It has been suggested [5.24] that the study of pion-nucleus elastic scattering in the DWIA approach can be used to settle some of the outstanding problems such as pion-nucleus optical potential, off-shell effects and medium modifications of the transition operator by avoiding nuclear wave function uncertainties.

5.4 Distorted Pion Waves

The pion optical wave functions obtained for $\pi^+ - {}^{16}O$ scattering are shown in Figs. 5.4–10. The $\nabla^2 \varrho(r)$ and $\nabla^2 \varrho^2(r)$ terms in the SMC potential tend to decrease the p-wave attraction as revealed by Fig. 5.4. What is the behavior of the pion wave functions obtained with the somewhat equivalent potentials, SMC and KB? This is illustrated by Figs. 5.5–8. These plots which show the real and imaginary parts of the pion wave functions obtained with the SMC and KB potentials for pion energies of 116 MeV and 170 MeV indicate, that the wave functions depict an increase in absorption in the nuclear region, as one approaches resonance. The SMC potential has a tendency to decrease the amount of absorption, as illustrated by the real parts of the wave functions (and $|u_l(r)|$ which are not shown). This is reflected in turn in the photopion cross sections with the SMC potential, that show an increase. The wave functions obtained with the gradient and local potentials

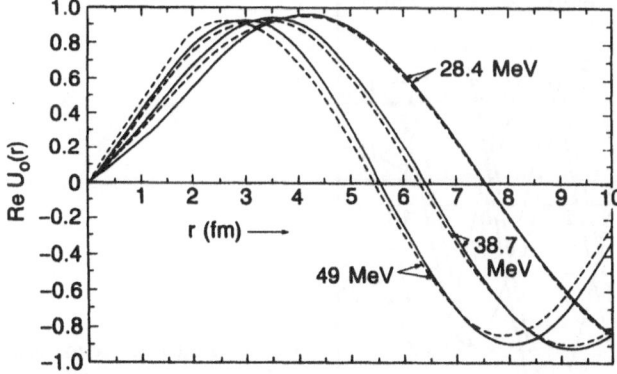

Fig. 5.4. Pion optical wave functions for $\pi^+ - {}^{16}O$ elastic scattering. The continuous and dashed curves are those obtained with and without $\nabla^2 \varrho(r)$ and $\nabla^2 \varrho^2(r)$ terms in the SMC potential

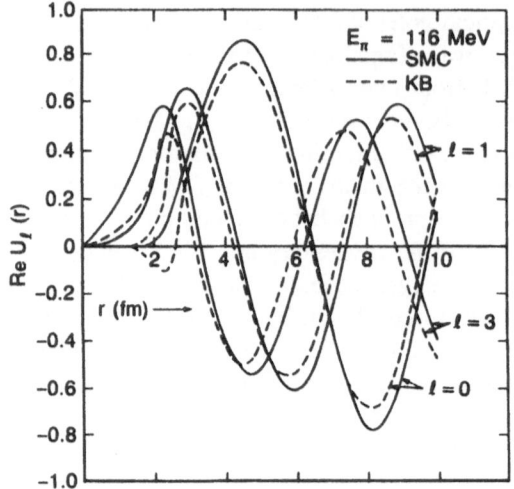

Fig. 5.5. Real parts of the distorted pion wave functions for $\pi^+ - {}^{16}$C scattering for the various pion partial waves for $E_\pi = 116$ MeV. The solid and dashed curves are those obtained with the SMC and KB potentials respectively

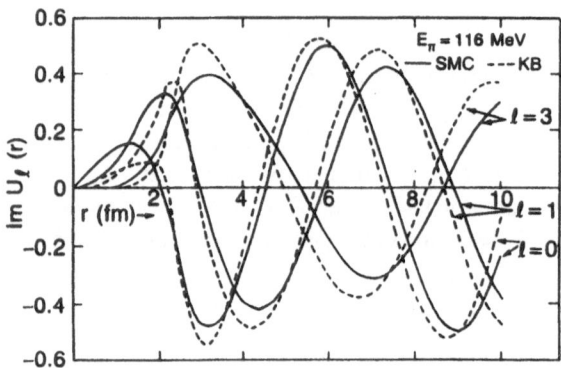

Fig. 5.6. Same as Fig. 5.5 for the imaginary parts of the wave functions

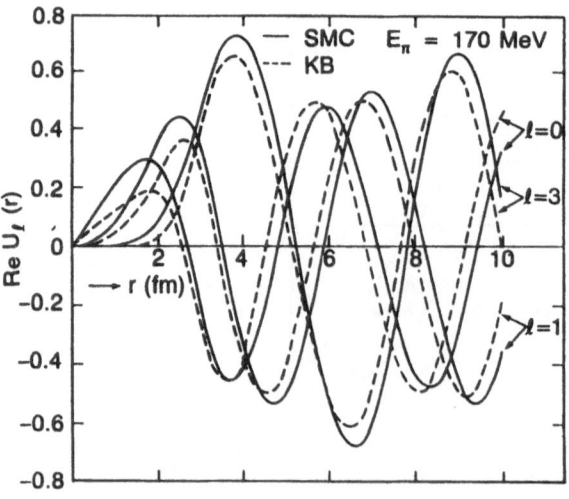

Fig. 5.7. Real parts of the pion optical wave functions for $E_\pi = 170$ MeV. See Fig. 5.5 for other details

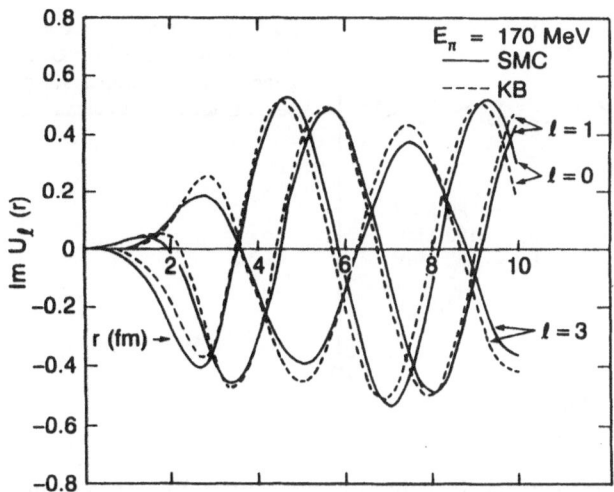

Fig. 5.8. Same as Fig. 5.7 for the imaginary parts of the optical wave functions

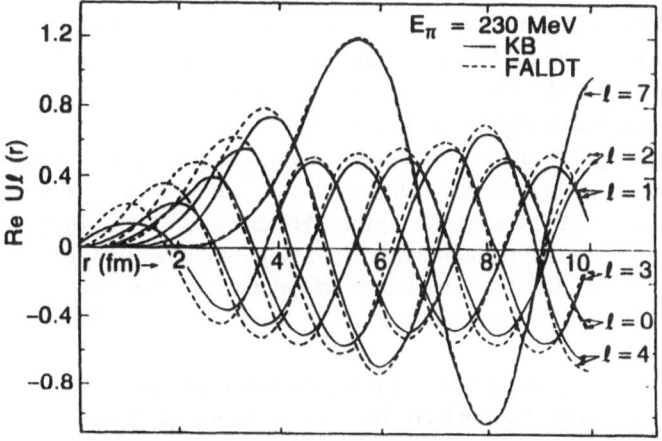

Fig. 5.9. Real parts of the pion wave functions for $\pi^+ - {}^{16}O$ scattering. The continuous curve corresponds to Kisslinger's potential and the dashed curve to Faldt's potential

are almost the same asymptotically. This can be inferred from Fig. 5.9, which depicts the real part of the wave function. The imaginary part shows a similar behavior. It is clear that there is not much of an absorption at 230 MeV. Again at energies well below resonance, the absorption is not appreciable. Around the resonance region, the reflection coefficient is very small for small l and never reaches unity even for higher l. This fact is illustrated in Fig. 5.10, wherein the wave functions for three different energies are plotted. This is in accordance with the fact that the imaginary part of the dominant parameter c_0 is maximum at resonance. This drastic reduction of the wave functions around resonance is responsible for bringing down the photopion cross sections in this energy range.

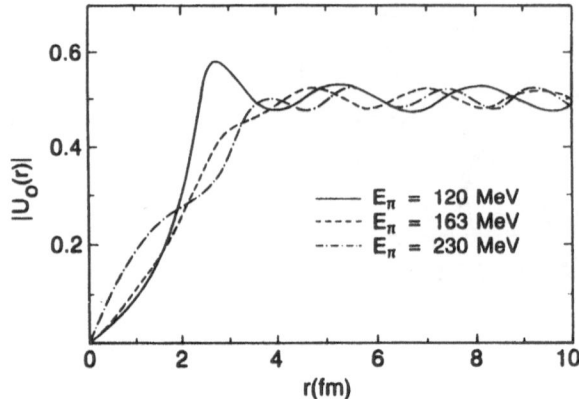

Fig. 5.10. The modulus of the pion wave functions for $\pi^+ - {}^{16}O$ scattering at various energies. KB potential is used in the calculation

Another problem is related to the fact that the elementary photoproduction operator as used in nuclear calculations contains implicitly all final state interactions between the produced pion and the active nucleon, while on the other hand the optical potential describes the interaction of the pion with the whole nucleus, i.e. including the active nucleon. There is clearly some double counting involved. Whether this is simply a $1/A$ effect which would be tolerable for nuclei with $A \geq 6$, or whether the situation is more serious (particularly at energies approaching the (3,3) resonance region) is an open question.

Finally, there are indications that the use of a phenomenological optical potential may be inadequate altogether for describing pion-nucleus final state interactions at energies approaching the resonance region. *Wittman* and *Mukhopadhyay* [5.25] obtain significantly better agreement with data at a pion energy of 116MeV if the (charged) pion photoproduction cross section is calculated using pion wave functions obtained by the Δ-hole appraoch instead of using optical model wave functions.

It has been repeatedly suggested [5.26,27] that pion wave functions obtained from a phenomenological optical model are inadequate in the (3,3) resonance region.

It has been pointed out that Δ-nucleus dynamics requires careful treatment, particularly near the resonance region, in order to obtain reliable results, and it has been argued that a phenomenological optical model is fundamentally unable to provide an adequate description of these effects.

6. Nuclear Wave Functions and Nuclear Transition Densities

The photopion cross section depends sensitively on nuclear structure [6.1,2]. The nuclear structure information required in nuclear photoproduction calculations consists, in principle, in the wave functions describing the initial and final nuclear states (between which the nuclear transition operator is evaluated). In practice, it is usually more convenient to express the nuclear structure input in terms of transition densities.

In the DWIA, the operators acting on the nuclear wave functions are the single nucleon operators τ, σ, p_i and p_f which appear in the transition operator in a variety of combinations. By writing the single nucleon photoproduction operator as

$$t = \sum_{i=1}^{N} A_i O_i \quad , \tag{6.1}$$

assuming that there are N different combination of nucleon operators O_i with the associated coefficients A_i, the nuclear transition amplitude

$$T = \left\langle J_f M_f \left| \Phi_\pi^*(r) \sum_{j=1}^{A} t_j\, e^{i\mathbf{k}\cdot\mathbf{r}} \right| J_i M_i \right\rangle \tag{6.2}$$

can be written in terms of the nuclear transition densities

$$\varrho_i(r) = \left\langle J_f M_f \left| \sum_j O_i^{(j)} \delta(\mathbf{r} - \mathbf{r}_j) \right| J_i M_i \right\rangle \quad . \tag{6.3}$$

Inserting $\varrho_i(r)$ in (6.2), we obtain

$$T = \sum_i \int \varrho_i(r)\, \Phi_\pi^*(r)\, e^{i\mathbf{k}\cdot\mathbf{r}} A_i\, d^3r \quad . \tag{6.4}$$

From the structure of the elementary operator, one finds that the nuclear information needed in pion photoproduction consists of the isospin and spin-isospin transition densities both with and without nucleon momenta, i.e., the transition densities required are those for $O_i = \tau$, $\tau\sigma$, τp_i, $\tau\sigma \cdot p_i$ and $\tau\sigma \cdot p_f$. More specifically, for a nuclear transition, the appropriate multipoles $\varrho^{LL'SJ}$ of these transition densities have to be found.

For the operator combinations τ and $\tau\sigma$ one has $L' = L$ and the associated transition densities are $\varrho^{LLOJ}(r) = \varrho^J(r)$ and $\varrho^{LL1J}(r) = \varrho^{LJ}(r)$, specifying the non-spin flip and spinflip densities, respectively. These are the only two densities needed if the nucleon momenta are not treated as operators, i.e. if they are e.g. replaced by the average values $p_i \rightarrow \langle p_i \rangle$ and $p_f \rightarrow \langle p_f \rangle$, an approximation that is often made. If the nucleon momenta are retained as operators, the additional densities $\varrho_{in}^{LL'1J}(r) = \varrho_{in}^{LJ}(r)$ associated with the initial nucleon momentum and $\varrho_f^{LL'1J}(r) = \varrho_f^{LJ}(r)$ associated with the final nucleon momentum are needed. In a shell model calculation, each of these nuclear transition densities is represented as a linear superposition of single particle transition densities [1.12]

$$\varrho_i^{LL'SJ}(r) = \sum_{ab} P_{ab}^J \varrho_{ab}^{LL'SJ}(r) \tag{6.5}$$

where the coefficients P_{ab}^J are identified as the reduced matrix elements of one body transition densities

$$P_{ab}^J = \langle J_f \parallel [a_a^+ \times a_b]_J \parallel J_i \rangle \tag{6.6}$$

which contain all the nuclear structure information.

The coefficients P_{ab}^J express the probability that a transition of multipolarity J occurs between given initial and final nuclear states by transfering a single nucleon from orbital state b to orbital state a. They are numbers depending only on the configuration mixing of wave functions of the initial and final nuclear states. *Singham* and *Tabakin* [1.12] tabulated the coefficients P_{ab}^J for some of the nuclear transitions that they have studied. They can also be generated using the Argonne National Laboratory (ANL) shell model code.

If it is assumed that nucleon transitions occur only within the $1p$ shell, then at most four coefficients are required for each transition. Most of the recent theoretical and experimental investigations were concerned with transitions between $1p$ shell nucleons and this assumption has been found to be a reasonable one [6.3] for the nuclear transitions considered. Since all nucleons in the $1p$ shell have the same parity, it is clear, however, that this assumption can be valid only for transitions without parity change. For transitions involving parity change, nucleon states outside the $1p$ shell necessarily contribute and in general a much larger set of coefficients P_{ab}^J has to be dealt with.

It is often convenient to express the nuclear structure information not in terms of the jj-coupling coefficients P_{ab}^J but through the LS-coupling coefficients Q_{LS}^J which express the relative weights with which a transition of multipolarity J between given initial and final nuclear states occurs through any of the allowed LS-channels. Values for S are restricted to $S = 0$ (non-spinflip channels) and $S = 1$ (spinflip channels). For nucleons restricted to the $1p$ shell, only $L = 0, 1, 2$ are allowed. The two sets of coefficients are related by

$$P_{ab}^J = \sum_{L,S} \hat{j}_a \hat{j}_b \hat{L} \hat{S} \begin{Bmatrix} 1 & \frac{1}{2} & j_a \\ 1 & \frac{1}{2} & j_b \\ L & S & J \end{Bmatrix} Q_{LS}^J \tag{6.7}$$

where the symbol $\hat{j}$ denotes $\sqrt{2j+1}$. For a $M1$ transition, e.g., the nuclear structure information is thus either given by the four coefficients P_{ab}^J or by Q_{10}^1, Q_{01}^1, Q_{11}^1 and Q_{21}^1.

At present no theoretical model is available which yields wave functions for a broad range of nuclear states from which the required transition densities can be determined reliably with adequate accuracy. Cohen–Kurath wave functions [6.3], which cover all nuclei within the $1p$-shell have been widely used. They are evaluated using two-body potential matrix elements which have been obtained by fitting selected energy levels throughout the $1p$-shell. Over the years, these wave functions have been applied with considerable success in a number of photoproduction calculations. Recently, however, several transitions have been studied in which calculations based on transition densities derived from Cohen–Kurath wave functions predict cross sections up to four times higher than measured in experiments. A number of examples are discussed in Chap. 7. It was found that the only reliable way to obtain transition densities which lead to cross sections in reasonable agreement with experimental results was to resort to some phenomenological approach by representing the nuclear states involved in a suitably restricted basis of particle -hole states and tightly constraining the free parameters by using experimental information from related reactions.

The strongest constraint is usually provided by inelastic electron scattering data. Here the transition densities are to lowest order

$$\varrho(r) = \langle J_f M_f | \varrho^{op}(r) | J_i M_i \rangle \quad ,$$

$$\boldsymbol{j}_c(r) = \langle J_f M_f | \boldsymbol{j}_c^{op}(r) | J_i M_i \rangle \quad ,$$

$$\boldsymbol{\mu}_s(r) = \langle J_f M_f | \boldsymbol{\mu}_s^{op}(r) | J_i M_i \rangle \quad ,$$

corresponding to the transition charge, convection current and magnetization density, respectively. They contain some combinations of nucleon operators which do not appear in the transition operator for photoproduction. In some cases, it is, however, possible to extract the components needed for photoproduction calculations. This is the case when some or all of the unwanted components are known to vanish or can be assumed to be small. Then a direct relationship between multipole densities required in photoproduction and the multipole densities measured in electron scattering can be established. This can be accomplished most easily by fitting the given electron scattering form factors with some parametric expressions as, e.g., suggested by the generalized Helm model [6.4–6]. The photoproduction transition densities already being expressed in terms of the parameters used in the fits are then immediately known. This approach has been successfully applied to a number of cases by *Überall* and his collaborators [6.7,8]. The generalized Helm model and its applications to nuclear pion photoproduction are discussed in Appendix A.

In general, it turns out, however, that constraining nuclear transition densities by inelastic electron scattering data in order to obtain reliable results in pion photoproduction calculation is only a necessary but not a sufficient condition. Other valuable constraints were found to be elastic electron scattering data, magnetic

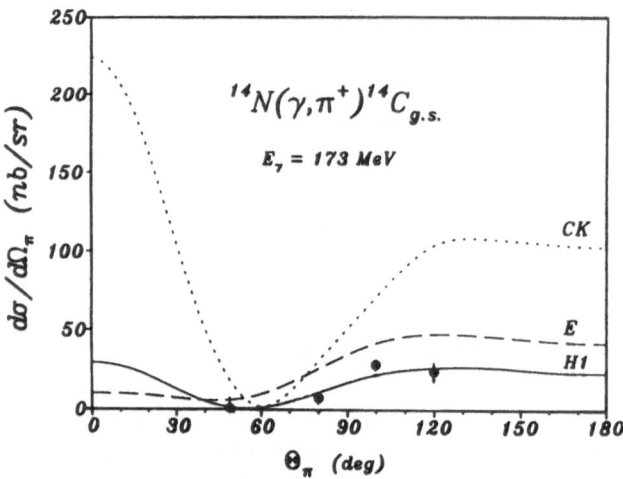

Fig. 6.1. Differential cross section for $^{14}N(\gamma, \pi^+)^{14}C_{g.s.}$ at $E_\gamma = 173$ MeV. Results of momentum space calculations using Cohen and Kurath (*CK*), MIT–NBS (*E*) and Huffman (*H*1) wave functions along with experimental data from [6.9]

dipole and electric quadrupole momenta, ft-values for β decay, and μ-capture data. The sensitivity of nuclear pion photoproduction cross sections to nuclear structure inputs is dramatically illustrated by some recent calculations [6.1,2] for the reactions $^{13}C(\gamma, \pi^+)^{13}B$ and $^{14}N(\gamma, \pi^+)^{14}C$.

In Fig. 6.1 which was taken from [6.9], the differential cross section for the reaction $^{14}N(\gamma, \pi^+)^{14}C$ at $E_\gamma = 173$ MeV is shown for three different sets of nuclear wave functions. The curve marked *CK* was obtained using Cohen–Kurath wave functions. These wave functions were found to predict the elastic $M1$ form factor as well as the magnetic and quadrupole momenta of ^{14}N correctly, but the value of the ^{14}C beta decay is predicted too small and more importantly, the inelastic $M1$ form factor for $^{14}N_{g.s.}(e, e')^{14}N^*(2.3\text{MeV})$ is predicted substantially too high in the relevant region of momentum transfer.

The curve marked *E*, calculated using Ennslin wave functions [6.10] shows a much better agreement with the data than the *CK* result. The wave functions used here provide agreement with the magnetic and quadrupole momenta and the log ft value. They also reproduce well the inelastic scattering form factor. However, the prediction of the elastic form factor is substantially too low.

The curve labeled *H*1 is seen to be in excellent agreement with the data. This calculation is based on the Huffman wave functions [6.11] which are obtained from an analysis of recent electron scattering data. These wave functions are designed to predict the magnetic moment correctly and to lead to a good description of both the elastic and the inelastic $M1$ form factors.

The resulting reduced density matrix elements Q_{LS}^J for the three types of wave functions are given in Table 6.1.

Several facts are noteworthy: the strong suppression of the normally dominant term Q_{01}^1, the large reduction of Q_{21}^1 obtained by using inelastic electron

Table 6.1. Reduced matrix elements of one-body transition densities in LS coupling scheme for the nucleon transition $^{14}N \rightarrow {}^{14}C_{g.s.}$

Q_{LS}^J	Cohen–Kurath	Ensslin	Huffman
Q_{10}^1	0.356	0.208	0.339
Q_{01}^1	−0.094	0.001	−0.003
Q_{01}^1	−0.188	−0.448	0.046
Q_{21}^1	0.792	0.482	0.434

scattering as a constraint, and the large magnitude and variability of the elements of the purely nonlocal $L = 1$ channels.

In conclusion, we wish to reiterate that the nuclear photopion cross section is very sensitive to nuclear wave functions and hence it can be used as a convenient probe for nuclear structure. Since the charged pion production excites selectively the spin and isospin states of nuclei, the photopion reactions can be used with advantage to investigate the structure of such unstable states which may not be accessible in other reactions.

7. Charged Pion Photoproduction

The study of charged pion photoproduction has attracted greater attention than the photoproduction of neutral pions for the simple reason that it is much easier to detect the charged pions. The recent advances in accelerator and detector technology have made it possible to measure the differential cross section with such a precision that it is possible now to examine the mechanism of charged pion production in nuclei and to investigate the influence of nuclear structure and the final state interaction. Specific nuclear transitions highlight the contributions of particular terms in the elementary photopion production amplitude and choosing those nuclear transitions wherein the dominant term in the elementary amplitude is suppressed, it is possible to examine closely the terms that are less important and contribute negligibly to the cross section for the elementary process. Available experimental data are analysed with a view to draw information on the nuclear structure, the effect of non-locality, the medium modification of the elementary operator and the role of delta. Most of the experiments that have been performed hitherto are at a gamma ray energy of about 200 MeV. Of late, experiments in the (3,3) resonance region are being conducted and they will throw light on the energy dependence of the reaction and the relative importance of the various terms in the elementary amplitude at different energies.

7.1 Total Cross Section Measurements

Reactions of the type $A(\gamma, \pi^{\pm})B$, involving nuclear transitions from well defined initial to final states, are of specific interest to us. In these cases, reliable theoretical calculations can be made and by comparison with the experimental results it is possible to draw inference on the mechanism of photopion production and also on nuclear structure.

The total cross sections for reactions of the type $A(\gamma, \pi^{\pm})B$ have been measured experimentally by identifying the residual nucleus B by some characteristic activity such as β or γ decay. The earliest measurement was done in 1958 by *Hughes* and *March* [2.20] for the reaction $^{11}B(\gamma, \pi^{-})\,^{11}C$ by identifying the final nucleus by its positron activity. Subsequently, there has been a continuous program of study of photopion production reactions of the type $A(\gamma, \pi^{+})B$. *Hummel* [2.21,23] and his coworkers have reported cross sections for the reactions $^{11}B(\gamma, \pi^{-})\,^{11}C$, $^{11}B(\gamma, \pi^{+})\,^{11}Be$, $^{16}O(\gamma, \pi^{+})\,^{16}N$ and $^{27}Al(\gamma, \pi^{+})\,^{27}Mg$, *Nydahl* and

Forkman [2.24] for the reactions ^{27}Al(γ, π^+) ^{27}Mg and ^{51}V(γ, π^+) ^{51}Ti, *Blomqvist* et al. [7.1] for the reactions ^{41}K(γ, π^+) ^{41}A, ^{27}Al(γ, π^+) ^{27}Mg and ^{65}Cu(γ, π^+) ^{65}Ni and *March* and *Walker* [2.22] for the reactions ^{60}Ni(γ, π^-) ^{60}Cu. It may be observed that in the experimental study of these reactions, the emitted pion is not detected and what one observes is the radioactivity of the final nucleus. So, the reported experimental cross sections correspond to a group of low-lying final nuclear states stable against nucleon emission. One can, in general, obtain information about these low-lying levels from nuclear spectroscopic tables and theoretical calculations are performed for nuclear transitions to these levels, accompanied by production of photopions. The sum of the partial cross sections, so calculated, has to be compared with the experimental results.

Such calculations have been performed earlier by *Laing* and *Moorhouse* [1.15] and *Devanathan* et al. [2.26–28] using the shell model description for the nucleus and a surface production model for the reaction mechanism. It is to be observed that the surface production model is purely a phenomenological model invoked to take into account in a rough way the effect of nuclear wave functions and the final state interaction of the outgoing pion with the final nucleus. Now it is possible to handle them in a rigorous way and they have been discussed in detail in the earlier chapters.

The measurement of total cross section by observing the radioactivity of the final nucleus generally includes many low-lying nuclear states which are stable against nucleon emission but it is possible to pick out reactions in which the final nucleus has only one state, stable against nucleon emission. One such reaction is ^{12}C(γ, π^-) ^{12}N in which the final nucleus corresponds to the ground state of ^{12}N(1^+) and this has been suggested as a good candidate by *Rao* et al. [2.28] for the detailed study in order to draw definite conlcusions regarding the mechanism of photopion production from nuclei.

7.2 Differential Cross Section

Although total cross section measurements of exclusive reactions leading to well-defined final nuclear states are somewhat favourable to the study of the various factors that go into the theoretical calculation, a more stringent test of the theory will come only from a comparison with differential cross section. Fortunately with the commissioning of high duty cycle electron accelerators and the construction of pion spectrometers, the differential cross section measurements have become feasible. The first differential cross section measurement has been reported by *Shoda* et al. [1.21] for the reaction ^{12}C(γ, π^+) ^{12}B$_{g.s.}$ and subsequently a wealth of data has become available for a variety of light nuclei. The data can be broadly classified into two groups — those reactions for which the Kroll–Ruderman (KR) term $\sigma \cdot \varepsilon$ gives the dominant contribution to the cross section and others for which the KR term is suppressed. Energy-wise also, the data can be classified — those with gamma ray energy above threshold for pion production but less than 200 MeV and those with gamma ray energy in the Δ resonance region. The accelerators at Sendai and Mainz are designed for energies less than 200 MeV and so all the measurements made in these laboratories are restricted to this energy. At Sendai, differential cross sections were measured at angles ranging from 30° to 150° in steps of 20° for reactions leading to the discrete states of the final nucleus. The final state could be identified by an accurate measurement of the energy of the detected pion. Several experiments on a variety of target nuclei have been performed from 1977 onwards and a wealth of data [1.21, 7.2–21] has been collected at E_γ = 200 MeV. These experiments have been very useful in checking the DWIA theory for photopion production and analysing the nuclear wave functions.

At Bates Laboratory, in the initial measurement of the differential cross sections, the pion spectrometer was kept at 45° or 90° and the differential cross sections were measured at various energies going beyond the (3,3) resonance region [2.59]. In this laboratory, experiments can be performed in the resonance region E_γ = 320 MeV. Already results [7.22] are available for selected target nuclei and they are of immense value in studying the medium modifications of the photopion production operator and the role of Δ in the photopion reaction at the resonance energy.

7.3 Discussion of Specific Reactions

It is not our intention to give an exhaustive account of all the reactions that have been studied hitherto. We wish to choose only typical cases which are of specific interest. The reactions $^{12}C(\gamma, \pi^+)^{12}B$, $^{12}C(\gamma, \pi^-)^{12}N$ and $^{16}O(\gamma, \pi^+)^{16}N$ involve target nuclei with closed shell or subshell, for which the nuclear wave functions are fairly well understood and these reactions have been investigated by various authors. They have been used as a test for the DWIA theory. In these reactions, the KR term gives the dominant contribution and hence the non-locality and the medium effects are not severe. We shall also consider reactions such as $^{14}N(\gamma, \pi^+)^{14}C_{g.s.}$ in which the KR term is suppressed. The reactions $^{13}C(\gamma, \pi^-)^{13}N_{g.s.}$ and $^{14}N(\gamma, \pi^+)^{14}C_{g.s.}$ are of unusual interest for a variety of reasons. Initially, there existed large discrepancies between theoretical predictions and experimental results. The intense theoretical effort prompted by these discrepancies has lead to a considerable refinement of the theoretical formalism.

Because of their importance, only the reactions listed above will be discussed in detail. For other reactions, the reader is referred to Table 7.1 wherefrom he can find the relevant references.

7.3.1 The Reactions $^{12}C(\gamma, \pi^-)^{12}N$ and $^{12}C(\gamma, \pi^+)^{12}B$

Several are the motivations for the study of charged pion photoproduction from ^{12}C

$$\gamma + ^{12}C \rightarrow \pi^- + ^{12}N_{g.s.} \tag{7.1}$$

Table 7.1. Charged pion photoproduction from light nuclei

S. No	Photopion reaction	Nuclear transition	References
1	$^3\text{He}(\gamma,\pi^+)^3\text{H}$	$\left(\frac{1}{2}^+,\frac{1}{2}\right) \rightarrow \left(\frac{1}{2}^+,\frac{1}{2}\right)$	1.9, 1.11
2	$^6\text{Li}(\gamma,\pi^+)^6\text{He(g.s.)}$	$(1^+,0) \rightarrow (0^+,1)$	7.4, 7.29
3	$^7\text{Li}(\gamma,\pi^-)^7\text{Be(g.s.)}$	$\left(\frac{3}{2}^-,\frac{1}{2}\right) \rightarrow \left(\frac{3}{2}^-,\frac{1}{2}\right)$	2.54, 7.29
4	$^9\text{Be}(\gamma,\pi^+)^9\text{Li(g.s.)}$	$\left(\frac{3}{2}^-,\frac{1}{2}\right) \rightarrow \left(\frac{3}{2}^-,\frac{1}{2}\right)$	7.5
5	$^{10}\text{B}(\gamma,\pi^+)^{10}\text{Be(g.s.)}$	$(3^+,0) \rightarrow (0^+,1)$	7.20, 7.29, 2.59
6	$^{10}\text{B}(\gamma,\pi^+)^{10}\text{Be}^*(3.37\,\text{MeV})$	$(3^+,0) \rightarrow (2^+,1)$	7.20
7	$^{10}\text{B}(\gamma,\pi^-)^{10}\text{C(g.s.)}$	$(3^+,0) \rightarrow (0^+,1)$	7.20
8	$^{12}\text{C}(\gamma,\pi^+)^{12}\text{B(g.s.)}$	$(0^+,0) \rightarrow (1^+,1)$	1.21, 2.38, 2.39, 4.21 7.2, 7.8, 7.23, 7.29
9	$^{12}\text{C}(\gamma,\pi^+)\text{B}^*(0.95\,\text{MeV})$	$(0^+,0) \rightarrow (2^+,1)$	1.21, 4.21, 7.2, 7.8, 7.29
10	$^{12}\text{C}(\gamma,\pi^+)\text{B}(4.302,$ $4.37,\ 4.521\,\text{MeV})$	$(0^+,0) \rightarrow (1^-,1),$ $(2^-,1),\ (4^-,1)$	7.2, 7.7, 7.8, 7.29
11	$^{12}\text{C}(\gamma,\pi^+)^{12}\text{N(g.s.)}$	$(0^+,0) \rightarrow (1^+,1)$	2.54, 7.23, 7.29
12	$^{13}\text{C}(\gamma,\pi^+)^{13}\text{B(g.s.)}$	$\left(\frac{1}{2}^-,\frac{1}{2}\right) \rightarrow \left(\frac{3}{2}^-,\frac{3}{2}\right)$	6.1, 7.3, 7.16, 7.34
13	$^{13}\text{C}(\gamma,\pi^-)^{13}\text{N(g.s.)}$	$\left(\frac{1}{2}^-,\frac{1}{2}\right) \rightarrow \left(\frac{1}{2}^-,\frac{1}{2}\right)$	4.21, 7.17
14	$^{14}\text{C}(\gamma,\pi^-)^{14}\text{N}(2.31\,\text{MeV})$	$(0^+,1) \rightarrow (0^+,1)$	1.10
15	$^{14}\text{N}(\gamma,\pi^+)^{14}\text{C(g.s.)}$	$(1^+,0) \rightarrow (0^+,1)$	4.21, 7.13, 7.19, 7.29
16	$^{14}\text{N}(\gamma,\pi^+)^{14}\text{C}^*(7.01\,\text{MeV})$	$(1^+,0) \rightarrow (2^+,1)$	7.9
17	$^{14}\text{N}(\gamma,\pi^+)^{14}\text{C}(8.32\,\text{MeV})$	$(1^+,0) \rightarrow (2^+,1)$	7.9
18	$^{14}\text{N}(\gamma,\pi^-)^{14}\text{O(g.s.)}$	$(1^+,0) \rightarrow (0^+,1)$	7.29
19	$^{15}\text{N}(\gamma,\pi^-)^{15}\text{O(g.s.)}$	$\left(\frac{1}{2}^-,\frac{1}{2}\right) \rightarrow \left(\frac{1}{2}^-,\frac{1}{2}\right)$	2.46, 7.14
20	$^{16}\text{O}(\gamma,\pi^+)^{16}\text{N(g.s.)}$ $0.12,\ 0.30,\ 0.40\,\text{MeV}$	$(0^+,0) \rightarrow (2^-,1),$ $(0^+,1),\ (3^-,1),\ (1^-,1)$	2.23, 2.41, 2.59, 7.6 7.22

$$\gamma + {}^{12}\text{C} \rightarrow \pi^+ + {}^{12}\text{B}_{\text{g.s.}} \ . \tag{7.2}$$

In the first place, the negative pion photoproduction leads to the ground state of $^{12}\text{N}(1^+)$ which is the only state stable against nucleon emission. In the case of π^+ production, the final state $^{12}\text{B}(1^+)$ can be singled out since it is well separated from the next excited state $^{12}\text{B}(2^+)$ by about 0.95MeV. Hence a direct comparison of theory with experiment is facilitated with a possibility of drawing definitive conclusions regarding the mechanism of photopion reactions. Besides a large body of information is available on ^{12}C to calibrate the nuclear structure inputs. The good quality data available for the $M1$ electron scattering to the analogous state at 15.1 MeV test the reliability of the nuclear wave functions.

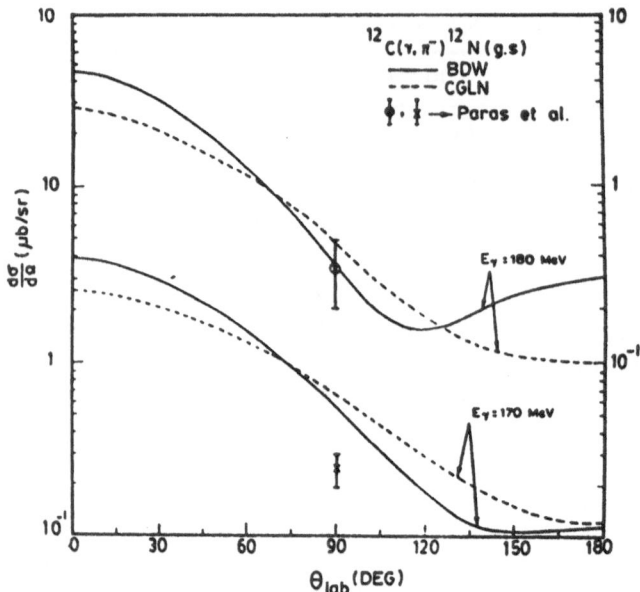

Fig. 7.1. Differential cross section for $^{12}C(\gamma, \pi^-)$ $^{12}N_{g.s.}$. The experimental data of *Paras* et al. [7.25] are compared with the theoretical calculations of *Girija* et al. [7.23] using BDW and CGLN amplitudes. The y-axis scale is marked on the left for $E_\gamma = 170$ MeV and on the right for $E_\gamma = 180$ MeV

Further it is possible to extract information regarding the iso-vector part of the pion-nucleus interaction since in these charged pion reactions (7.1) and (7.2) the pion interacts with the unstable excited nucleus of isospin $T = 1$.

These reactions have been theoretically investigated by *Singham* and *Tabakin* [2.39], *Nagl* and *Überall* [2.38] and *Girija* et al. [7.23]. Girija et al. have used the $1p$ generalized wave functions of *Hirooka* et al. [7.24] for the initial and final nuclear states, the CGLN and BDW amplitudes for the elementary process and the pion-nucleus optical potential of *Stricker* et al. [2.37] for the final state interaction.

The differential cross sections obtained by them are given in Figs. 7.1 and 7.2 along with the available experimental results. There is reasonable agreement between theory and experiment [7.25] for reaction (7.1) at $E_\gamma = 180$ MeV but the theoretical cross section is somewhat higher at $E_\gamma = 170$ MeV. For reaction (7.2), there is good agreement between theory and the experimental data of *Shoda* et al. [1.21]. It would be instructive to do experiments at higher energies so that the DWIA theory can be tested at resonance energies where the medium effects and the delta propagation are likely to be important.

7.3.2 The Reaction $^{16}O(\gamma, \pi^+)$ ^{16}N

In this reaction, there are four low lying closely spaced states 2^-, 0^-, 3^-, 1^- ($E_x = 0$, 0.12, 0.30 and 0.40 MeV) of ^{16}O that contribute. These four states are the only states that are stable against nucleon emission and hence contribute to

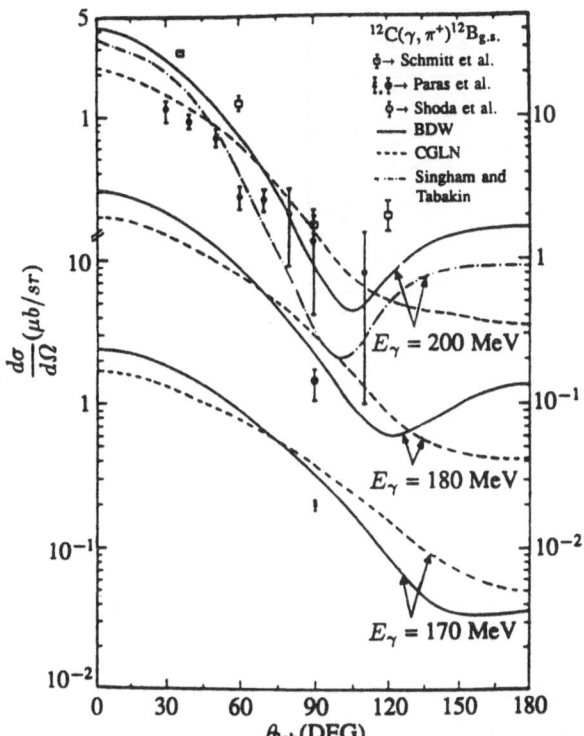

Fig. 7.2. Differential cross section for $^{12}C(\gamma, \pi^+)^{12}B_{g.s.}$ along with the available experimental data. The y-axis scale for E_γ = 170 and 200 MeV are to the left and the scale is marked on the right for E_γ = 180 MeV

the total cross section measurement of *Meyer* et al. [2.23] by the observation of β-activity. Even in the differential cross section measurement by *Shoda* et al. [7.6], the four states are not resolved.

This reaction has been investigated by *Girija* and *Devanathan* [2.41] using the CGLN and BDW single nucleon amplitudes and the SMC optical potential for the final state pion-nucleus interaction. The single nucleon amplitude involves momentum operators and the effect of these terms are investigated by replacing the momentum by a gradient operator. As one would expect, the momentum dependent terms are of minor importance at threshold energies since the KR term dominates at low energies. At higher energies, the momentum dependent terms dominate and hence at resonance, they may have more influence on the cross section. In Fig. 7.3, the experimental data of *Shoda* et al. [7.6] at E_γ = 198MeV are compared with the theoretical calculations of *Girija* and *Devanathan* [2.41]. The calculated partial cross sections to the discrete final states J^π = $2^-, 0^-, 3^-, 1^-$ of ^{16}N are also given along with their sum which is to be compared with the experimentally measured cross section. Since the 0^- state has contributed negligible cross section, it is not marked in Fig. 7.3. The momentum operator when replaced by a gradient operator gives a better agreement with experimental data.

Besides the experimental data at E_γ = 198 MeV, data were also available [2.59] at two fixed angles (θ = 45° and 90°) as a function of energy going beyond

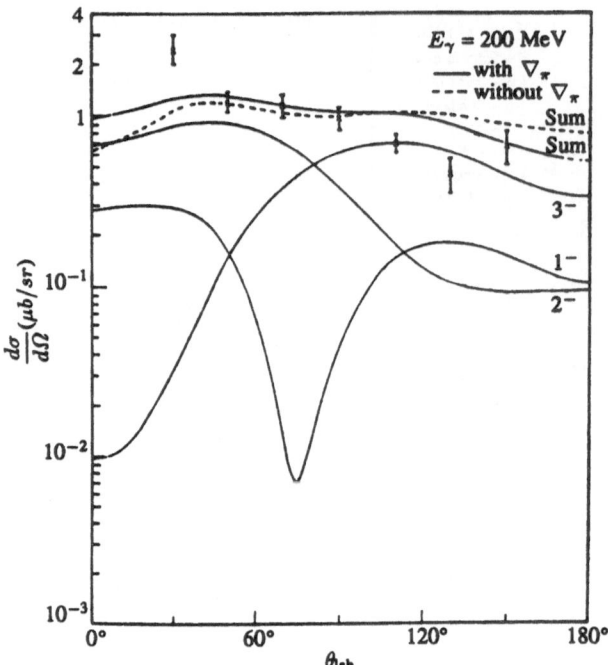

Fig. 7.3. Differential cross section for $^{16}O(\gamma, \pi^+)\,^{16}N(2^-, 0^-, 3^-, 1^-)$ at $E_\gamma = 200$ MeV. Sum denotes the sum of the partial cross sections leading to 2^-, 0^-, 3^- and 1^-. The experimental data are taken from [7.6]

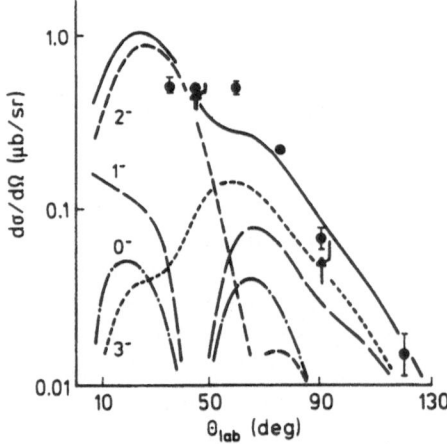

Fig. 7.4. Differential cross section for $^{16}O(\gamma, \pi^+)$ $^{16}N(2^-, 0^-, 3^-, 1^-)$ at $E_\gamma = 320$ MeV. Experimental data are taken from [7.22] and theoretical calculations are due to *Eramzhyan* [7.26]

the (3,3) resonance. These have large experimental uncertainties. Recently, a full angular distribution measurement has been reported at the resonance energy by *Yamazaki* et al. [7.22]. The momentum space DWIA calculation of *Eramzhyan* et al. [7.26] compare favourably with the experimental data (see Fig. 7.4). The agreement in this energy range is rather unexpected and it may be due to a fortuitous cancellation of disagreements between experiment and theory for the individual transitions. Or the calculated partial cross sections may be approximately correct but the medium effects may be relatively small for the dominant transitions. So, a more detailed theoretical study is required before coming to any definite conclusion.

7.3.3 The Reaction $^{14}\text{N}(\gamma, \pi^+)\,^{14}\text{C}_{\text{g.s.}}$

The $M1$ transition involved in this reaction is of outstanding interest because of the near-vanishing of the allowed Gamow–Teller matrix element [2.39]. The KR term which normally dominates in the photoproduction operator is suppressed here so that terms which are usually small may become relatively large and possibly even dominant. This makes the $A = 14$ system an attractive test laboratory to investigate terms in the transition operators which are normally difficult to study, and to test the adequacy of the theoretical models used in the calculations. It is expected that medium effects would be amplified in cases like the $A = 14$ systems so that inadequacies of the DWIA might show up more pronounced here than in other cases.

Because the cross sections for photoproduction on ^{14}N are unusually low, measurements of differential cross sections were practical only after both high duty factor electron accelerators and pion spectrometers with a large solid angle and good momentum and angular resolution became available. These conditions were not satisfied until very recently. At Bates as well as at Mainz, pion spectrometers with the necessary specifications now exist and they have already been used to produce differential cross sections of excellent quality for π^+ production on ^{14}N at photon energies of 173, 200, 260 and 320 MeV.

Initial calculations of cross sections for the process $^{14}\text{N}(\gamma, \pi^+)\,^{14}\text{C}_{\text{g.s.}}$ showed great sensitivity to the nuclear wave functions used [2.39,43]. A reduction of the total cross sections by more than a factor of 2 was found when switching from CK [6.3] to the phenomenological MIT-NBS wave functions [7.27].

Recently additional sets of empirical wave functions were derived by *Huffman* et al. [6.11] which were fitted to both elastic and inelastic electron scattering data. In contrast, the MIT-NBS wave functions were fitted only to inelastic data and they do not reproduce the recent elastic data [6.11], whereas the CK wave functions are consistent only with the elastic but not with the inelastic data.

In Fig. 6.1 (taken from [6.9]) theoretical differential cross sections for $^{14}\text{C}(\gamma, \pi^+)\,^{14}\text{C}$ at $E = 173$ MeV, calculated with three different sets of wave functions, are shown together with the data of *Rohrich* et al. [6.9]. It is apparent that the $H1$ wave functions, giving the best fit to the data, are clearly to be preferred over the MIT-NBS wave functions which lead to the result labeled E. The inadequacy of the Cohen–Kurath wave functions (CK) is apparent. The data taken at Bates at a photon energy of 200 MeV and the result of the calculation using the $H1$ wave functions (solid line) are shown in Fig. 7.5.

Calculations using $H1$ wave functions thus appear to be in excellent agreement with the data at both 173 and 200 MeV. It has to be kept in mind, however

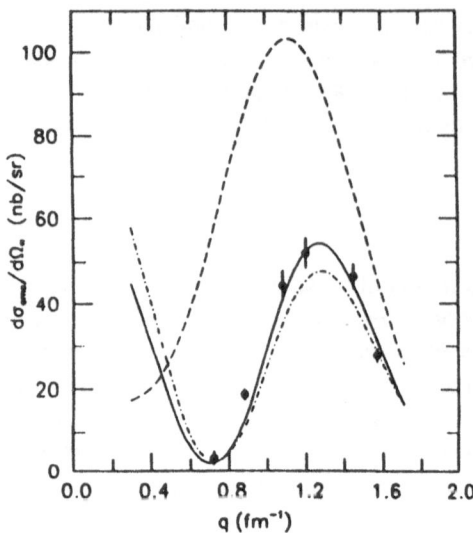

Fig. 7.5. Differential cross section for ^{14}N(γ, π^+) ^{14}C$_{g.s.}$ at E_γ = 200 MeV. *Solid line*: full BL operator; *dot-dashed line*: BL operator without Δ term; *dashed line*: KR term only. Data are taken from [4.22]

that Coulomb effects and corrections to the elementary amplitude [7.28], which affect cross sections at the 20 % level at the energies considered, were not included in the calculations.

Despite these uncertainties, it is clear that the gap that had existed earlier between ^{14}N photoproduction data and theoretical predictions has now essentially been closed, at least in the low energy region. The improvement in the theoretical predictions was, however, not only due to the use of more tightly constrained nuclear wavefunctions (i.e. the switch to the $H1$ set), but in a large measure also due to the fact that nonlocal effects associated with the propagators in the transition amplitude were fully taken into account. All calculations shown in Figs. 6.1 and 7.5 were obtained with the DWIA momentum space code of *Tiator* and *Wright* [2.42] which explicitly accounts for all nonlocalities. The importance of nonlocal effects can be seen in Fig. 7.6 where the nonlocal result (solid line) is compared to the result obtained in the local approximation (dotted line). In both

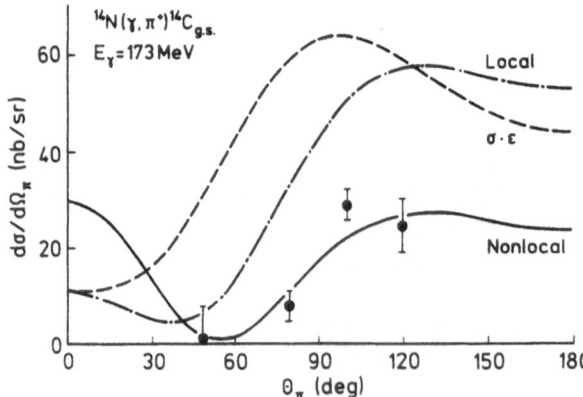

Fig. 7.6. Differetial cross section for ^{14}N(γ, π^+) ^{14}C$_{g.s.}$ at E_γ = 173 MeV. Data from [7.13] and calculations using $H1$ wave functions. *Dashed line*: KR term only; *dotted line*: full operator, local approximation; *solid line*: full non-local calculation

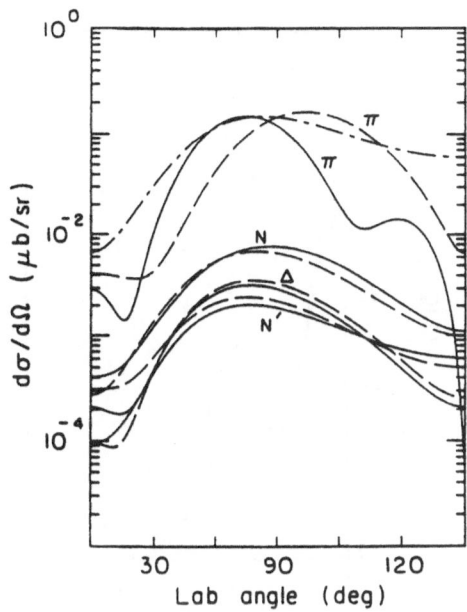

Fig. 7.7. Contributions of individual Feynman diagrams of BL amplitude to the differential cross section for $^{14}N(\gamma, \pi^+)$ $^{14}C_{g.s.}$ at $T_\pi =$ 50 MeV [2.46]. *Solid curves*: non-local calculation; *dashed curves*: local limit; *dot-dashed curve*: KR term only

Figs. 7.5 and 7.6, one can observe the strong destructive interference between the KR term and the momentum dependent terms which reduces the cross section from the dashed lines (KR result) to the full lines if the full transition operator is used.

In Fig. 7.7 (taken from [2.46]), the contribution of individual diagrams to the cross section at a pion kinetic energy of 50 MeV is shown. Both the local (dashed curves) and the nonlocal results (solid curves) are indicated. It is apparent that the large nonlocal and interference effects are mainly due to the pion pole term which is the only momentum-dependent term that leads to a contribution which is comparable in magnitude to that of the (here strongly suppressed) KR term (shown as the dot-dashed curve). In Fig. 7.5 the cross section obtained with the Δ term removed (dashed-dotted line) is also shown. The significance of the Δ term can thus be assumed to be small at energies at 200 MeV and below.

Very recently measurements of π^+ photoproduction on ^{14}N have been extended to higher energies, first by Stoler and Seneviratne (as quoted in [7.28]) to $E_\gamma = 260$ MeV and then by *Teng* et al. [4.12] to the Δ region ($E_\gamma = 320$ MeV). *Wittman* and *Mukhopadhyay* [7.28] used their upgraded elementary transition operator to carry out a theoretical analysis at these two energies as well as at $E_\gamma = 200$ MeV.

The theoretical formalism used by *Wittman* and *Mukhopadhyay* [7.28] is similar to that of *Tiator* and *Wright* [2.42]. The calculations are also carried out in momentum space using $H1$ nuclear wave functions and the SMC pion-nucleus optical potential [2.37]. In addition, however, the Coulomb part of the pion-nucleus final state interaction is explicitly accounted for, pion wave functions generated by the Δ-hole approach are used as an alternative to the optical

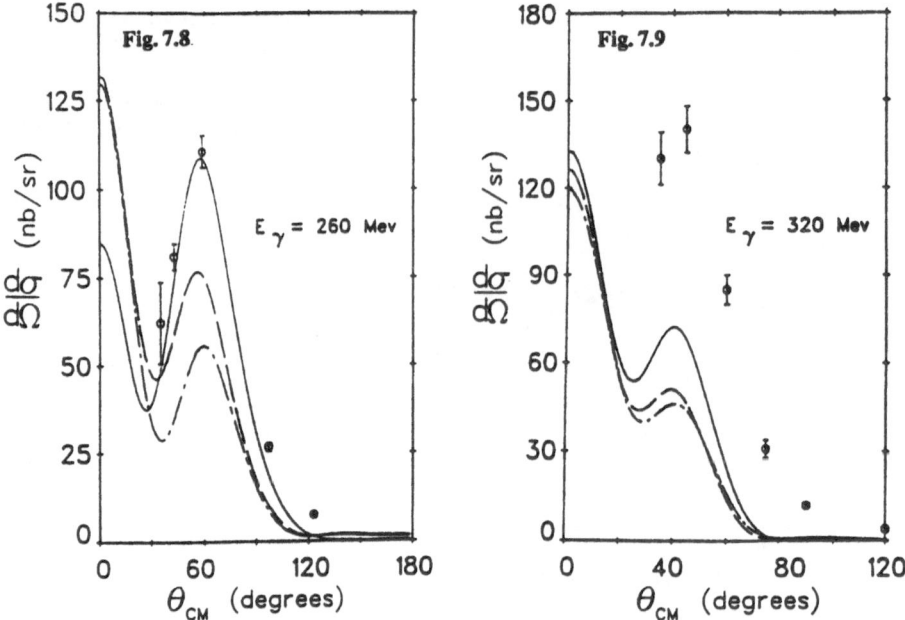

Fig. 7.8. Differential cross section for $^{14}N(\gamma, \pi^+)\,^{14}C_{g.s.}$ at $E_\gamma = 260\,\text{MeV}$. Calculations by *Wittman* and *Mukhopadhyay* [7.28]; *solid line*: full amplitude with pionic FSI from Δ-hole approach; *dashed line*: full amplitude with FSI using SMC optical potential; *dot-dashed line*: BL amplitude with FSI using SMC optical potential

Fig. 7.9. Differential cross section for $^{14}N(\gamma, \pi^+)\,^{14}C_{g.s.}$ at $E_\gamma = 320\,\text{MeV}$. For details, refer the figure caption of Fig. 7.8

potential wave functions, and a number of corrections — unitarization of BL amplitude, addition of u-channel, Δ, and t-channel ω^0 contributions, re-introduction of second $\gamma N\Delta$ gauge coupling — to the standard BL transition operator [2.45], are included.

The results obtained by Wittman and Mukhopadhyay at 260 MeV together with the preliminary data points of Stoler and Seneviratne are shown in Fig. 7.8. Here the effect of the corrections to the transition operator manifests itself as the difference between the dashed curve (all corrections included) and the dash-dotted curve (BL amplitude). Coulomb effects are negligible in this case. The corrections are seen to lead to a net increase in the peak value of the cross section. However, even the corrected curves falls far short of most of the data points. Very much improved agreement with the data is obtained if the final-state pion distortions obtained from the SMC optical potential [2.37] are replaced by those calculated with the Δ-hole approach.

At 320 MeV, the corrections to the transition operator lead to only a small effect as seen by the small difference between the dash-dotted line (BL amplitude) and the dashed line (corrected BL amplitude) in Fig. 7.9. Here again the peak value in the cross section curve is substantially shifted upward by the switch to Δ-hole pion wave functions, but at this energy the result is a factor of two short of what is required by the data points of *Teng* et al. [4.12]. Wittman and Mukhopadhyay speculate that the remaining discrepancy may be attributed partially to inadequacies of the nuclear wave functions used and partially to the fact that contributions from multi-step processes (mainly coherent production with subsequent charge exchange) have not been accounted for.

Since the charged pion production cross sections are anomalously low for ^{14}N and coherent production rates are quite substantial in the resonance region, it is concievable that in the present case the two-step process contributes significantly. It is unlikely, on the other hand, that inadequacies of the nuclear wave function set $H1$ are sufficient to explain the large gap between theoretical predictions and data at this energy. Electron scattering from factors calculated with the $H1$ wave functions do fall off too quickly at high q values as compared to the experimental curves. However, a good fit is obtained for momentum transfers up to about $q = 1.7\,\text{fm}^{-1}$, which, as pointed out by *Teng* et al. [4.12], corresponds in the present case to an angular range up to about 70°. The $H1$ wave functions should thus be adequate over most of the angular range in which data points are available, certainly in the region in which the data peak lies.

Theoretical results at 320 MeV were also obtained by *Teng* et al. [4.12], using the code of *Tiator* and *Wright* [2.42]. Their nonlocal calculation agrees with that of [7.28]. Noteworthy is the result [4.12] showing the cross section obtained if the Δ term in the transition operator is deleted. Comparison of this result with that of full nonlocal calculation indicates that in the case considered here the entire cross section in the peak region is due to the Δ term. This unusually strong Δ dependence has two important implications:

1) It indicates that outright elimination of the Δ term from the transition operator as is frequently advocated is not likely to be the correct approach, since in the present case it would drastically increase the already large gap between theory and data.
2) Medium modifications of the Δ term in the transition operator which were found to be significant in coherent π^0 production may play an important role also in the present case and may account for some of the discrepancies between theory and data.

A satisfactory theoretical description of charged pion photoproduction on ^{14}N appears now to be possible for photon energies from threshold up to 260 MeV. Nonlocal effects were found to be important over this whole energy range. Up to at least 200 MeV, pion wave functions obtained from the phenomenological optical potential constrained by pionic atoms and pion scattering data appear to be adequate. However, even at these energies a strong sensitivity to phase shift equivalent potentials was noted [7.29]. At 260 MeV, agreement with data can only be achieved if pion wave functions obtained from a Δ-hole calculation are used. There appears to be no need however, to account for medium effects at this energy. In other words, the DWIA appears to be adequate at least up to 260 MeV.

In the resonance region, on the other hand, serious discrepancies between DWIA results and data are found. At 320 MeV, corrections to the elementary amplitude and the use of improved pion wave functions bring the theoretical results only within a factor of two of the data. Within the DWIA itself, there appears to be little room left for improvement. The transition operator, pion wave functions and nuclear transition densities are well constrained. It thus appears that the 320 MeV data of *Teng* et al. [4.12] provide the first strong indication of a significant breakdown of the DWIA in charged pion photoproduction at the resonance energy.

7.3.4 The Reaction $^{13}C(\gamma, \pi^+)^{13}B_{g.s.}$

For the reaction

$$\gamma + {}^{13}C\left(J, T = \tfrac{1}{2}, \tfrac{1}{2}\right) \longrightarrow {}^{13}B_{g.s.}\left(J, T = \tfrac{3}{2}, \tfrac{3}{2}\right) + \pi^+$$

the magnitude of the cross sections suggests that the KR term is not suppressed as in the previous example but dominates. Nevertheless, large discrepancies between theory and experiment were found here also. So far, experiments for this reaction were limited to the low energy region ($\tau_\pi < 50$MeV). In this energy range, as can be inferred from the discussions of Chap. 6, both transition multipoles involved ($M1$ and $E2$) are expected to have little sensitivity to Δ effects and only the $M1$ transition is expected to be affected (but only moderately) by nonlocal effects. Thus no great theoretical uncertainties should exist with respect to the elementary operator. This qualitative assessment is confirmed by the results obtained by *Toker* and *Tabakin* [2.46].

In similar cases, local DWIA calculations using standard ingredients (BL operator or CGLN amplitudes and standard p-shell model wavefunctions) have generally achieved good agreement with data. In the present case, however, using the same approach, the predicted cross sections were uniformly much higher than the data. The differential cross section for $E_\gamma = 194$ MeV obtained by *Toker* and *Tabakin* [2.46] using CK wave functions [6.3] are shown in Fig. 7.10, where they are compared with the data obtained at the Tohoku linac by *Shoda* et al. [7.3]. It is evident that nonlocal effects change the cross section only insignificantly (dash-dotted line vs. solid line marked PV), and that even if the Δ term is dropped completely the resulting change in the cross section is much too small to explain the discrepancy with the data. The theoretical uncertainty associated with the PS and PV versions of the transition operator is indicated by the difference between the two solid lines. Other results [7.30] also using CK wave functions are also all substantially higher than the data, and in addition show a disturbingly large variation among themselves. It turns out that with Cohen–Kurath wavefunctions neither elastic nor inelastic electron scattering data are correctly reproduced. Both the elastic $M1$ form factor [7.31] and the inelastic transverse form factor for the transition to the 15.11 MeV level [7.3,31] are substantially overestimated for momentum transfers between 1 and 2 fm^{-1}.

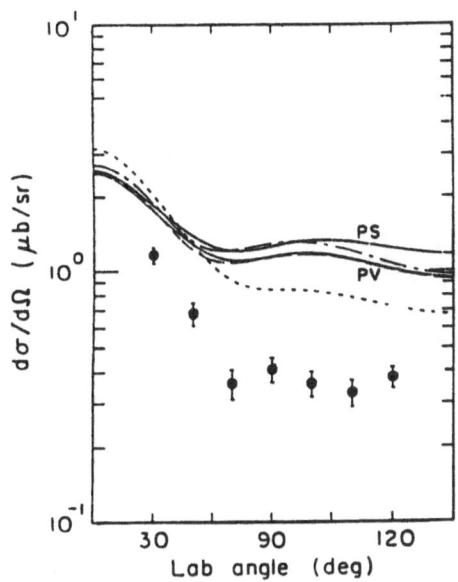

Fig. 7.10. Differential cross section for ^{13}C(γ, π^+) ^{13}B$_{g.s.}$ at $T_\pi = 43$ MeV. Data are from [7.3] and calculations from [2.46]. *Solid line*: nonlocal results for PS and PV pion-nucleon couplings; *dotted curve*: result for PV with Δ diagram omitted; *dot-dashed curve*: local results

Assuming the simple p-shell picture was too restrictive to adequately describe the nuclear structure in the present case, *Sato, Koshigiri* and *Ohtsubo* (as quoted in [7.3]) proceeded to improve the nuclear wave functions by incorporating core polarization effects, leading to sizeable admixtures of higher configurations into the p shell. This approach significantly lowered the predicted inelastic electron scattering form factors. Cross sections for pion photoproduction are similarly improved (see dashed line in Fig. 7.11). The solid lines in Fig. 7.11 indicate the result obtained if CK wave functions are replaced by Hauge–Maripuu (HM) wave functions [7.32]. In Fig. 7.11, the $M1$ and $E2$ contributions are indicated separately.

Singham [7.31] used a different approach for improving the nuclear structure input. He showed that it is possible to dramatically improve the agreement for both elastic and inelastic electron scattering form factors with configuration mixing entirely restricted to the $1p$-shell. He used as constraints on the nuclear wave functions not only properties of the ground state and the 15.11 MeV 1^+ state of ^{13}C but weak and electromagnetic data involving the analogues of these states. *Singham* [6.1] found that using the resulting ground state wave function I and either one of the two excited state wave functions A and B in pion photoproduction calculations yields cross sections which are considerably closer to the data than the results based on the CK wave functions. The set IA shown in Fig. 7.12 as solid line leads to the best result. This calculation was based on a localized BL elementary operator and pion wave functions obtained by using the latest SMC pion optical potential [2.37].

The sensitivity of the photoproduction cross sections to pion distortions is demonstrated by the dot-dashed curve in Fig. 7.12 which was obtained by using the 1979 version of the SMC potential [2.37]. Even though the latest version

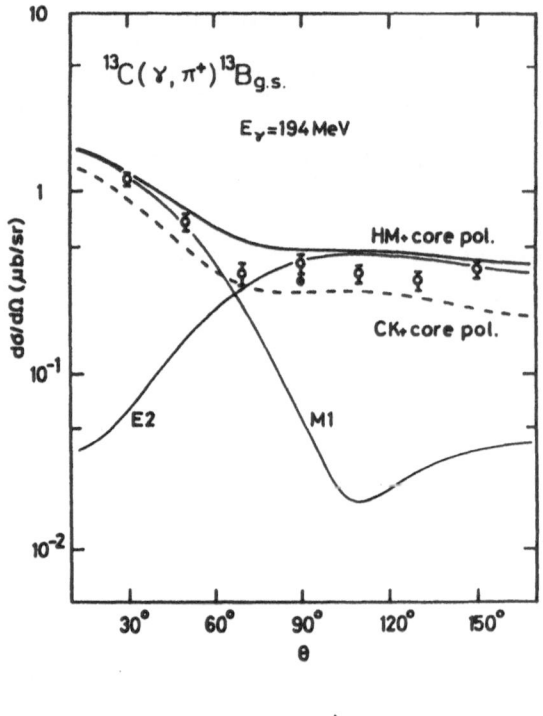

Fig. 7.11. Differential cross section for $^{13}C(\gamma, \pi^+)\,^{13}B_{g.s.}$. Data are from [7.3] and [7.34] (*solid circle*)

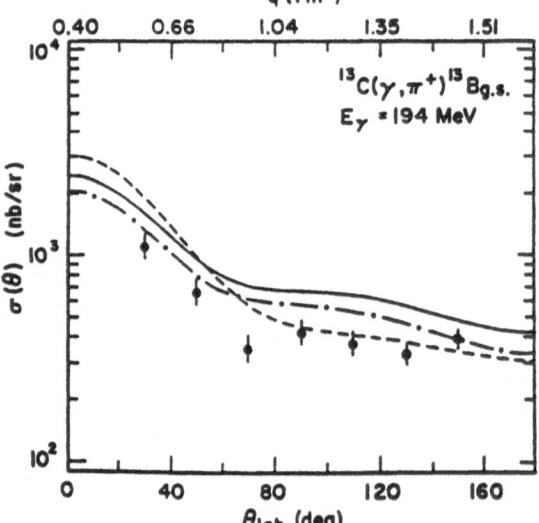

Fig. 7.12.
Differential cross section for $^{13}C(\gamma, \pi^+)\,^{13}B_{g.s.}$ at $E_\gamma = 194\,\text{MeV}$. Comparison of experimental data [7.3] with theoretical estimates of *Singham* [6.1]. *Solid line*: with full BL operator with optical potential (A); *dash-dot line*: with full BL operator with optical potential (B); *dashed curve*: without Δ term with optical potential (A)

of the potential is assumed to be more accurate (and is shown by Singham to give better agreement with elastic scattering at 47.3 MeV), the earlier version of the potential yields the better agreement with the photoproduction data. This contradictory situation is probably just due to the fact that the nuclear wave functions used, while clearly being superior to the CK wave functions, are by no means the last word. There are indications that future improvements to the wave functions (use of Woods–Saxon wave functions [7.33] and accounting for core polarization effects) may lead to additional reductions in the theoretical cross sections to the point where the more reliable optical potential will indeed lead to better agreement with the data.

The effect of deleting the Δ term is indicated in Fig. 7.12 by comparing the dashed curve with the solid curve.

Singham carried out his calculations in the local approximation. However, since the reaction under consideration is predominantly a local one as is evident from Fig. 7.10, a full nonlocal calculation is unlikely to produce any dramatic change in the differential cross sections.

7.3.5 The Reaction $^{13}\text{C}(\gamma, \pi^-)\,^{13}\text{N}_{\text{g.s.}}$

This is another example of a reaction for which anomalously low measured cross sections were encountered. The reaction involves isobaric analog states with $J_i = J_f = 1/2$, $T_i = T_f = 1/2$. It is interesting for several reasons. The allowed transition multipolarities are $E0$ and $M1$. Charged pion photoproduction cross sections are usually dominated by the non-resonant spinflip $M1$ contribution (the KR term). From elastic magnetic electron scattering, it is known [7.33] that for ^{13}C, the $M1$ form factor has a deep minimum at $q = 1.04\,\text{fm}^{-1}$. *Tiator* and *Wright* [2.42] have shown that for photoproduction the minimum shifts only slightly (to $q = 1.01\,\text{fm}^{-1}$). Thus by selecting kinematics in the vicinity of this value for the photoproduction experiment $^{13}\text{C}(\gamma, \pi^-)\,^{13}\text{N}_{\text{g.s.}}$ the $E0$ part of the reaction (which is known to be dominated by the delta resonance even down to relatively low pion energies [2.42]), can be studied separately. This means that here one is presented with an opportunity to investigate the delta contribution to photoproduction over an extended range of energies, including the low energy tail where the delta contribution is usually strongly dominated by the Born terms. In addition, this reaction has shown strongly enhanced sensitivity to nonlocality effects, both through the resonance term and the non-resonant part of the transition operator.

The initial measurements of differential cross sections [7.34] were taken at 90° for $T_\pi = 17$, 29 and 42 MeV, corresponding to momentum transfers in the range of $0.9 - 1.1$ fm^{-1}. These values placed the measurements near the minimum of the magnetic form factor. At the time the data were taken they were in sharper disagreement with theoretical predictions than any previous photopion measurements. In the meantime, the gap between theory and experiment has closed considerably but some puzzling discrepancies still remain.

Initial shell model calculations of the differential cross sections at 90° as a function of pion energy by *Singham, Tabakin* and *Dytman* (as quoted in [7.43]) and by *Maleki* [7.30], both using CK wave functions, differed from the experimental results by factors of about 20 and 4, respectively. Calculations based on the Helm model [7.34] using parameters obtained by fitting to the $M1$ elastic electron scattering data yielded results close to and slightly below those of Maleki.

A subsequent calculation by *Tiator* [7.35] was based on phenomenological wave functions which were obtained by using nuclear magnetic moments and beta decay and magnetic electron scattering data to constrain the values of the reduced one-body density matrix elements in terms of which cross sections are expressed. This calculation achieved much better agreement with the data, although like other calculations previously done, it did not consider nonlocality effects. *Tiator* and *Wright* [2.42] using their momentum space code, found that nonlocal effects also play an important role but did not lead to a better agreement with the data.

Recently angular distributions have been measured at fixed energies. In the first experiment [7.36], carried out at the NIKHEF – Amsterdam linac, cross sections were measured at 65°, 100° and 125° for a pion kinetic energy of 48 MeV. The data points [7.36] are shown in Fig. 7.13. Even though these values are an order of magnitude higher than the result of *LeRose* [7.34] at 90° and 42 MeV (shown by the dashed error bar in Fig. 7.13), calculations based on CK wave functions are still on average a factor of four too high with respect to the new data [7.36].

The most recent calculations by Tiator and Wright using the phenomenological nuclear transition densities of *Tiator* [7.35] agree significantly better with the new data. The local calculation has the correct shape but is a factor of two too high (dashed curve in Fig. 7.13). The nonlocal calculation [2.42] (solid curve in Fig. 7.13) shows the expected improvement over the local results only for the backward angles (100° and 125°), whereas in the forward directions nonlocality effects push the cross section higher, increasing the discrepancy with the data point at 65°.

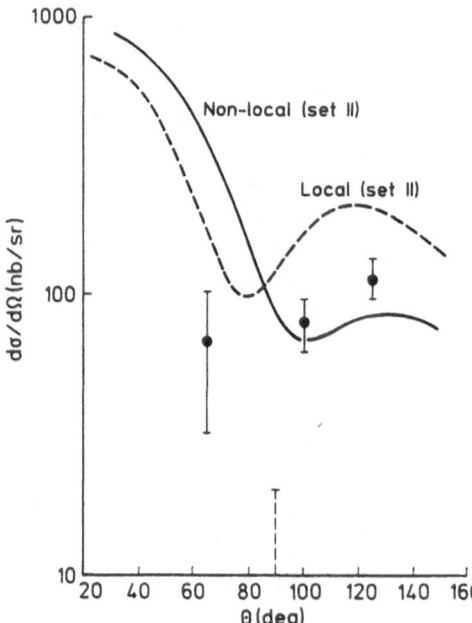

Fig. 7.13. Differential cross section for $^{13}C(\gamma, \pi^-)\,^{13}N_{g.s.}$ at $T_\pi = 48$ MeV. Data are from [7.18] and the vertical hatched line is due to *LeRose* [7.34] and the local and non-local calculations are due to *Tiator* and *Wright* [2.42]

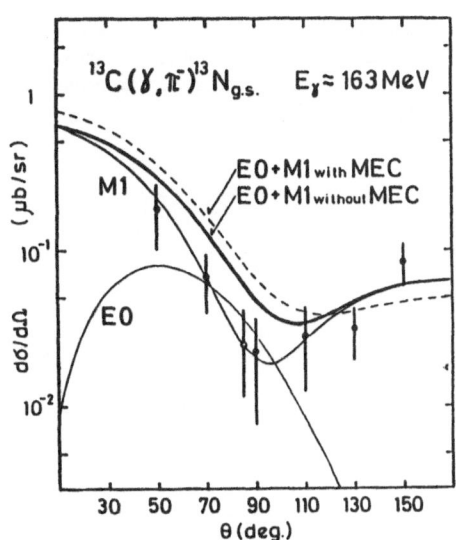

Fig. 7.14. Differential cross section for $^{13}C(\gamma, \pi^-)^{13}N_{g.s.}$ at $E_\gamma = 163$ MeV. Experimental data are due to *Shoda* et al. [7.17] (*solid circles*) and due to *LeRose* et al.[7.34] (*open circles*) and the effect of meson exchange current (MBC) is depicted

In the most recent experiment, carried out at the Tohuku linac by *Shoda* et al. [7.17], differential cross sections were measured between 50° and 150° at a photon energy of 163 MeV, corresponding to a pion kinetic energy of 19 MeV. The 90° measurement at 17 MeV of [7.34] is in excellent agreement with these results. In Figs. 7.14 and 7.15 the experimental data are compared with two non-local calculations both using the BL operator and pion wave functions based on the SMC optical potential. Ohtsubo et al. used Hauge and Maripuu wave-functions [7.32]. They considered in their calculation core polarization effects using parameters for the residual interaction determined by fitting to elastic electron scattering data. The thick solid curve in Fig. 7.14 shows the result of the calculation. The thin curves indicate the contributions due to the $E0$ and $M1$ transitions, respectively. If meson exchange currents are included, the theoretical cross sections are shifted away from the data (see dashed line in Fig. 7.14). The effect is however small. The data are well described by the $M1$ component alone, suggesting a strong suppression of the $E0$ component.

Results calculated by Wright using the formalism of [2.42] are shown in Fig. 7.15. The thick solid line shows again the full nonlocal result, the thin solid lines the contributions of the $E0$ and $M1$ components separately. The dashed lines give the associated results if the BL amplitude without the Δ term is used. If the Δ term is retained the data are reasonably well described (slightly underestimated) by the $M1$ component alone. The $E0$ contribution is here significantly larger than that obtained by Ohtsubo et al. As a result, the theoretical prediction here overestimates the data not only near the minimum but also in the forward directions. The best fit is obtained if the $E0$ component is suppressed by about an order of magnitude. Good agreement with the data is also obtained if both the $E0$ and $M1$ components are retained, but the Δ term is dropped.

From the preceding discussion, it is apparent that for this reaction there are at present still significant discrepancies between theory and data despite

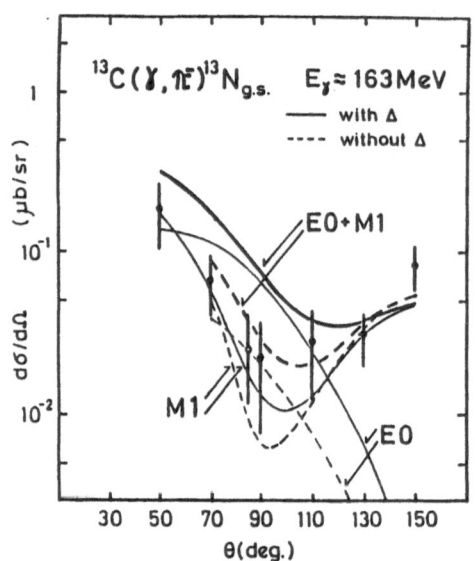

Fig. 7.15. Non-local calculation of Wright with (*solid curves*) and without (*dashed curves*) Δ contribution along with the experimental data (same as in Fig. 7.14)

the fact that recent calculations have been based on tightly constrained nuclear wave functions, and that nonlocal effects have been fully accounted for in the calculations. Most of the observed discrepancies appear to be due to the $E0$ transition.

The best agreement with the data is generally obtained if the $E0$ component is strongly reduced. Removing the Δ term from the BL operator was also found to lead to improved agreement with the data. It should be noted that these two observations are not independent, but are in fact closely connected. The $E0$ transition was found to be dominated by the Δ term [2.42], whereas in the $M1$ transition the Δ contribution is small [2.42]. Thus removing the Δ term from the BL operator and dropping the contribution due to the $E0$ transition are nearly equivalent.

Since for the $E0$ transition, no nuclear structure uncertainties exist [2.42] (at least within the $1p$ shell configuration space) and optical model wave functions are thought to be reliable at low energies, inadequacies in the transition operator are the most likely source for the discrepancies between theory and data. Significant uncertainties that exist with respect to the elementary operator in the present case are the role of medium effects and possibly the contribution of one of the corrections to the operator discussed by *Wittman* and *Mukhopadhyay* [7.28] (e.g. the contribution of the ω^0 in the t-channel).

The possibility that a two-step process, consisting of coherent π^0 production with subsequent charge exchange, leading to a significant contribution can also not be ruled out. Since the cross sections for the process under consideration are unusually small (about 25 nb/Sr) at 90° for a photon energy of 167 MeV, i.e., about three orders of magnitude lower than the cross section for coherent production (under the same conditions), it is quite conceivable that here the two-step process may be able to successfully compete with the direct process.

7.4 Concluding Remarks

The charged pion photoproduction is dominated by the Kroll–Ruderman term at threshold and hence the spin-flip nature of the interaction makes the charged pion photoproduction an interesting probe of nuclear structure complementary to electron scattering. At low energies the reaction dynamics are well understood and the DWIA theory is fairly successful. Hence the charged pion photoproduction reaction can be used to investigate the nuclear structure at low energies. Once the nuclear structure is understood, the reaction can be used to study the dynamics at higher energies, especially at resonance when the Δ channel becomes dominant. So, for a complete knowledge of this process, we require the experimental differential cross section for a wider range of energy starting from the pion threshold and going beyond the (3,3) resonance. At Bates, experiments have been done recently at higher energies and further experiments are being planned.

Differential cross section measurements have recently been made at resonance energy for the reactions $^{10}B(\gamma, \pi^+)\,^{10}Be_{g.s.}$, $^{14}N(\gamma, \pi^+)\,^{14}C_{g.s.}$ and $^{16}O(\gamma, \pi^+)$ $^{16}N(2^-, 0^-, 3^-, 1^-)$. More experiments are likely to be performed in the near future and it is hoped that they will clarify many of the existing anomalies and ambiguities.

Even at lower energies, some problems do exist. The reactions $^{13}C(\gamma, \pi^-)$ $^{13}N_{g.s.}$ and $^{15}N(\gamma, \pi^-)\,^{15}O_{g.s.}$ recieve contributions from both $E0$ and $M1$ transitions and the experimental data available at $E_\gamma \simeq 170$ MeV have revealed certain puzzles. *Bernstein* [1.10] has suggested the investigation of $^{14}C(\gamma, \pi^-)\,^{14}N(2.31$ MeV) which is a pure $E0$ transition since it corresponds to $0^+ \rightarrow 0^+$ nuclear matrix element. It would be worthwhile to look for some exotic nuclear transitions which will throw light on the relatively smaller terms in the photopion reactions when the dominant terms are suppressed.

8. Neutral Pion Photoproduction

The neutral pion photoproduction can take place either coherently or incoherently in a nucleus whereas the charged pion photoproduction can take place only incoherently. Only the spin-independent part of the elementary amplitude contributes for neutral pion photoproduction from a spin zero target nucleus. Since the Kroll–Ruderman term which is dominant in the case of charged pion production is absent, the neutral pion photoproduction takes place dominantly through the resonance term and hence it is amenable for treatment by an isobar doorway model or by an isobar-hole formalism besides the DWIA method. In this chapter, we shall discuss these methods and review the theoretical and experimental status on this subject.

8.1 An Overview

The coherent neutral pion photoproduction is of particular interest for the following reasons.

1) The coherent π^0 photoproduction can take place throughout the volume of the nucleus unlike the charged pion production for which the nuclear transition densities are peaked near the surface. So, the coherent π^0 production is best suited to investigate the production and propagation of the pion well in the interior of the nucleus.

2) For the coherent π^0 production on closed shell nuclei such as ^{4}He, ^{12}C, ^{16}O and ^{40}Ca, the nuclear wave function uncertainties are minimal since the process involves the nuclear density distributions which are better known without much of ambiguities.

3) For the spin zero target nucleus, only the spin-independent part of the elementary (γ, π^0) amplitude contributes and hence the coherent π^0 production can be used with advantage to distinguish between the different elementary photopion amplitudes which are tested to yield correctly the π^0 cross sections for the elementary process. In other words, the coherent π^0 production from spin-zero nuclei yields information regarding the finer details of the elementary amplitude just as the polarization measurements.

4) The Kroll–Ruderman term $\sigma \cdot \varepsilon$ which is dominant for charged pion photoproduction especially at threshold is absent in the case of neutral pion

photoproduction and hence the neutral pion production takes place predominantly through the $\Delta(1232)$ resonance. So, the coherent π^0 photoproduction is amenable for detailed study through the isobar-doorway model of *Kisslinger* and *Wang* [8.1–4] and through the microscopic isobar-hole approach of *Koch* and *Moniz* [8.5–7].

In the literature, it is found that three different approaches are used in the theoretical study of the coherent π^0 photoproduction from nuclei.

1) The Distorted Wave Impulse Approximation (DWIA) with the elementary amplitude of *Chew* et al. (CGLN) [2.15] and *Berends* et al. (BDW) [2.30] has been used by *Girija* et al. [8.8] and *Kamalov* and *Kaipov* [8.9] whereas *Boffi* and *Mirando* [8.10] have used the elementary amplitude of *Blomqvist* and *Laget* [2.45] in their DWIA calculations. Their results indicate that the coherent π^0 production from nuclei can well be described in the framework of DWIA which has been quite successful in the study of charged pion photoproduction. It may be recalled that *Saunders* [2.29] made the earliest attempt to study the coherent π^0 production from ^{12}C, ^{40}Ca and ^{208}Pb using the DWIA method but due to lack of reliable pion-nucleus optical potentials at that time, he reported cross sections which were much lower than the experimental data available then. This has been misconstrued by some authors as the failure of the DWIA theory to explain the coherent π^0 production from nuclei and thereby stimulated other approaches to this problem.

2) The isobar-doorway model (IDM) of *Kisslinger* and *Wang* [8.1,2] has been used successfully by *Saharia* and *Woloshyn* [8.4] for coherent π^0 photoproduction. It is a phenomenological model using essentially the π^0 production in the resonance channel. The medium effects and multiple rescattering effects are introduced in the development of the nuclear T-matrix in terms of the nucleon t-matrix. The parameters that occur in this model are fitted to yield the elastic π nucleus scattering data.

3) The microscopic isobar-hole $(\Delta - h)$ approach has been used by *Koch* and *Moniz* [8.5,6] for the study of the coherent (γ, π^0) reaction. It is claimed that several dynamical aspects of pion-nucleus interactions are handled in an essentially exact manner within the framework of the nuclear shell model. These include the usual multiple scattering effects, nucleon binding effects, isobar propagation and Pauli blocking. More complicated mechanisms such as pion absorption are treated phenomenologically through a spreading interaction of the isobar.

Although both the isobar-doorway model and the isobar-hole formalism [8.5] rest on the dominant feature of the production of isobars in the reaction channel, the isobar-doorway model involves the parametrisation of the pion-nucleus transition matrix with a resonant form whereas the isobar-hole formalism provides a vehicle for a microscopic calculation with only certain 'higher order' processes such as absorption treated approximately.

Before entering into a detailed discussion on the above theoretical methods, we shall briefly review the available experimental data base.

The experimental data on coherent π^0 photoproduction from nuclei are sparse [8.11–17] because of the inherent experimental difficulties in detecting the neutral pion and disentangling the contributions of the coherent process from the cross sections leading to excited nuclear bound states and to continuum states. Most of the experiments on π^0 photoproduction on medium nuclei ($A > 4$) have used the tip of the bremsstrahlung spectrum and the detection of π^0 decay photons near the minimum opening angle to isolate coherent processes. In the recent experiments at Bonn, *Arends* et al.[8.16] have developed an alternative approach of detecting π^0 with the simultaneous observation of the final nuclear system using an active target.

The early experiments on coherent π^0 production on spin zero target nuclei were performed at National Bureau of Standards [8.11] and they were mainly designed to observe the nuclear matter distribution since a simple-minded calculation directly relates the cross section to $|F(K)|^2$, K being the momentum transfer to the target nucleus and $F(K)$, the Fourier transform of the nuclear density distribution.

At the MIT synchrotron, experiments on ^{12}C and ^{40}Ca were performed by *Davidson* [8.12] in 1959 and since the energy resolution was 20MeV, it was likely that the data might have been contaminated with some incoherent events. The Davidson data has been subsequently renormalized to take into account changes in the elementary $\gamma P \rightarrow \pi^0 p$ cross section and this has been reported in Fig. 8 of [8.6]. A slight enhancement is observed in the corrected data. Also available are the data on ^{4}He$(\gamma, \pi^0)^4$He from *Staples* [8.13]. Recently, the Davidson experiment has been repeated by the Boston University group [8.15] with the MIT Bates Linear accelerator and further experimental data with improved energy resolution are expected in the near future.

The recent data [8.16] from the 500 MeV synchrotron at Bonn on ^{12}C have been obtained using a tagged photon beam with an energy resolution of 10 MeV. Also the total π^0 cross section (inclusive) on a wide variety of targets has been reported over the energy range $220 - 450$ MeV and an A-dependent $\sigma_A \propto A^{0.66}$ has been established [8.18]. Recently at Saclay [8.17], the coherent π^0 production on carbon has been measured near threshold with an energy resolution of 1 MeV using the tagged photon technique and the data compares favourably well with the DWIA calculations of *Boffi* and *Mirando* [8.10].

In the following pages, we shall briefly discuss the three theoretical approaches to the problem of coherent π^0 photoproduction from nuclei and assess their comparative success in explaining the available experimental data.

8.2 The DWIA Method

The nuclear transition amplitude for the coherent photoproduction of π^0 is given by [8.8]

$$T = \sum_j t_j = \sum_j \phi^*(q, r_j) \, t_s(j) \, \exp(\mathrm{i}\, k \cdot r_j) \tag{8.1}$$

where the index j runs over the A nucleons. The distorted pion wave function $\phi(q,r)$ takes into account the final state interaction of the outgoing pion with the nucleus and the incident photon is represented by a plane wave. The single nucleon transition amplitude t_s denotes the spin-independent part L of the elementary amplitude and it is of the general form

$$t_s = L = C\, q \cdot k \times \epsilon \tag{8.2}$$

where q denotes the momentum of the outgoing pion, k the momentum and ϵ the polarization vector of the incident photon. The factor C can be assumed from anyone of the commonly used elementary amplitudes given by *Chew* et al. (CGLN) [2.15], *Berends* et al. (BDW) [2.30], and *Blomqvist* and *Laget* (BL) [2.45].

The pion and photon wave functions are expanded into partial waves

$$\phi(q,r) = 4\pi \sum_{l',m'} (\mathrm{i})^{l'} g_{l'}(qr)\, Y_{l'}^{m'}(\hat{r})\, Y_{l'}^{m'*}(\hat{q}) \tag{8.3}$$

$$\exp(\mathrm{i}\, q \cdot r) = 4\pi \sum_{l,m} (\mathrm{i})^{l} j_l(kr)\, Y_l^m(\hat{r})\, Y_l^{m*}(\hat{k}) \quad . \tag{8.4}$$

The radial part of the pion wave function of $g_{l'}(qr)$ is obtained by solving the Klein–Gordon equation with a suitable pion-nucleus optical potential and $j_l(kr)$ is the spherical Bessel function. The pion momentum in the transition amplitude is replaced by the gradient operator ∇ and this is expected to take care of the non-locality and the modification of the transition operator due to many body interactions. Using the gradient formula

$$\begin{aligned}
\nabla_\mu \left(g_{l'}(qr)\, Y_{l'}^{m'}(\hat{r}) \right) = \sum_L & \left(\frac{2l'+1}{2L+1} \right)^{1/2} C(l'1L,m'\mu)\, C(l'1L,00) \\
& \times Y_L^{m'+\mu}(\hat{r}) \Big\{ \delta_{L,l'+1} \left(D_-(l')\, g_{l'}(qr) \right) \\
& + \delta_{L,l'-1} \left(D_+(l')\, g_{l'}(qr) \right) \Big\}
\end{aligned} \tag{8.5}$$

and the partial wave expansions (8.3) and (8.4), the nuclear transition matrix element for coherent π^0 photoproduction from the spin zero nucleus can be written as

$$Q = \langle \phi_0 | T | \phi_0 \rangle \quad , \tag{8.6}$$

where ϕ_0 denotes the wave function of the ground state nucleus. Equivalently, the nuclear matrix element can be written using the nuclear density distribution $\varrho(r)$ and the single nucleon transition amplitude $t(r)$ defined in (8.1) in the distorted wave approximation

$$Q = \int t(r)\, \varrho(r)\, d^3r \quad . \tag{8.7}$$

After some angular momentum algebra for which we follow the notations and conventions of *Rose* [4.1], we finally obtain

$$Q = C(4\pi)^2 \sqrt{2} \sum_{l,l'} (i)^{l-l'} (-1)^l S \int \varrho(r) R(l', l, r) r^2 dr \tag{8.8}$$

with

$$S = \frac{(-1)^l}{\sqrt{4\pi}} \sum_{\mu} C(111, 0\mu) C(1l', l, \mu, -\mu) k\varepsilon_1^\mu Y_{l'}^{-\mu}(\hat{q}) \tag{8.9}$$

and

$$R(l', l, r) = (2l' + 1)^{1/2} C(l', 1l, 00) \left\{ \delta_{l,l'+1} \left(D_-(l') g_{l'}(qr) \right)^* \right.$$
$$\left. + \delta_{l,l'-1} \left(D_+(l') g_{l'}(qr) \right)^* \right\} j_l(kr) \quad . \tag{8.10}$$

Expressions (8.8–10) are identical with (6, 7, 8) of *Girija* et al. (hereafter referred to as (GDNU) [8.8].

In the limiting case of plane waves for pions (PWIA), the nuclear matrix element reduces to

$$Q = 4\pi L \int j_0(Kr) \varrho(r) r^2 dr \tag{8.11}$$

where L is defined in (8.2) and $K = k - q$ is the momentum transfer.

In [8.8], a gaussian density distribution

$$\varrho(r) = \frac{A}{\pi^{3/2} a^3} \exp\left(-\frac{r^2}{a^2}\right) \tag{8.12}$$

has been assumed for ^{4}He with size parameter $a = 1.24$ fm and a Fermi distribution

$$\varrho(r) = \frac{\varrho_0}{1 + \exp\left(\frac{r-c}{t}\right)} \tag{8.13}$$

with the normalization constant

$$\varrho_0 = \frac{3A}{4\pi c^2 \left(1 + \frac{\pi^2 t^2}{c^2}\right)} \tag{8.14}$$

for ^{12}C, ^{16}O and ^{40}Ca. The following values for the size parameter c and the thickness parameter $T = t \ln 3$ are consistent with the electron scattering data.

$c = 2.3$ fm , $\quad T = 2.0$ fm $\quad$ for $\quad ^{12}$C ,

$c = 2.6$ fm , $\quad T = 2.0$ fm $\quad$ for $\quad ^{16}$O ,

$c = 3.51$ fm , $\quad T = 2.47$ fm $\quad$ for $\quad ^{40}$Ca .

An alternative method of obtaining the nuclear density distribution $\varrho(r)$ is to use the shell model description for the nucleus and the oscillator wave functions for the orbital states

$$\varrho(r) = N_1\langle j_0(Kr)\rangle_{1s} + N_2\langle j_0(Kr)\rangle_{1p} + N_3\langle j_0(Kr)\rangle_{1d} + N_4\langle j_0(Kr)\rangle_{2s}$$

$$(8.15)$$

where N_1, N_2, N_3 and N_4 are the total number of nucleons in the $1s$, $1p$, $1d$ and $2s$ shells and $\langle j_0(Kr)\rangle$ denotes the expectation value of $j_0(Kr)$ taken between the corresponding orbital states. For example,

$$\langle j_0(Kr)\rangle_{1s} = \int u_{1s}^*(r)\, j_0(Kr)\, u_{1s}(r)\, r^2\, dr \quad , \tag{8.16}$$

where $u_{1s}(r)$ denotes the oscillator wave function for the $1s$ state. The density distribution obtained in this way are consistent with the density distribution given by (8.13).

The DWIA treatment outlined above follows closely the work of GDNU [8.8]. They have made extensive calculations on the coherent π^0 photoproduction from ^{4}He, ^{12}C, ^{16}O and ^{40}Ca in the energy range $E_\gamma = 170 - 380\,\text{MeV}$ and compared them with the experimental data [8.12–18] and also with the other theoretical calculations done earlier using the isobar-doorway model. The following observations emerge from their study.

1) The PWIA calculations yield larger cross sections when compared to DWIA in the γ-ray energy range $230 - 380\,\text{MeV}$, see Fig. 8.1.

2) At lower energies, the DWIA cross sections are higher than the PWIA cross sections, see Fig. 8.1.

3) The CGLN amplitude yields higher cross sections than the BDW amplitude at γ-ray energies below $300\,\text{MeV}$ but at higher energies, the situation is reversed. Figures 8.1 and 8.2 illustrate this point. It is found that the spin-independent amplitude L which alone contributes to the coherent π^0 production differs in CGLN and BDW amplitudes although both the amplitudes yield correctly the cross sections for the elementary process $\gamma + \text{p} \rightarrow \pi^0 + \text{p}$. This is because the sum of the squares of the spin-dependent amplitude K and the spin-independent amplitude L are equal in both cases, i.e.,

$$|K|^2_{\text{CGLN}} + |L|^2_{\text{CGLN}} = |K|^2_{\text{BDW}} + |L|^2_{\text{BDW}}$$

although

$$|L|^2_{\text{CGLN}} > |L|^2_{\text{BDW}} \qquad \text{for} \quad E_\gamma < 320\,\text{MeV}$$

$$|L|^2_{\text{CGLN}} \approx |L|^2_{\text{BDW}} \qquad \text{for} \quad E_\gamma \approx 320\,\text{MeV}$$

$$|L|^2_{\text{CGLN}} < |L|^2_{\text{BDW}} \qquad \text{for} \quad E_\gamma > 320\,\text{MeV} \quad .$$

So, we find that a measurement of the cross section alone for the elementary process cannot distinguish between these two amplitudes. Polarization data will be a better test to make a choice between them.

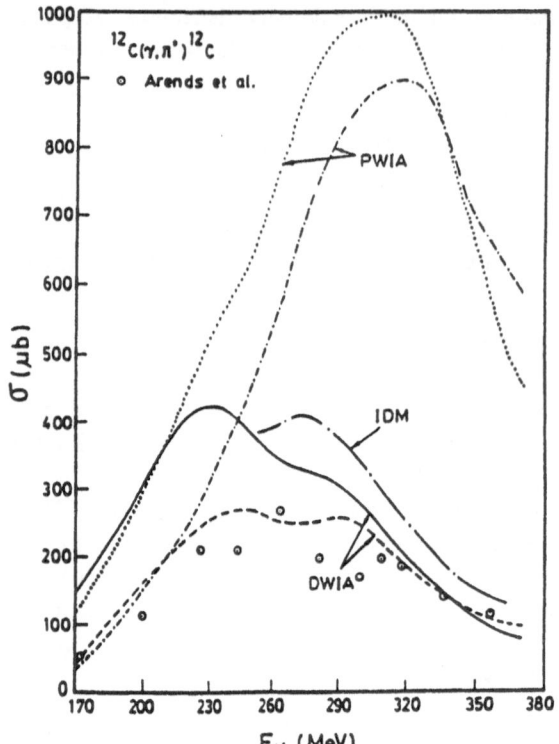

Fig. 8.1. Total cross sections for $^{12}C(\gamma, \pi^0)\,^{12}C$. PWIA and DWIA calculations [8.8] with CGLN (*dotted* and *solid lines*) and BDW (*dashed-dotted* and *dashed curves*) amplitudes along with IDM calculations [8.4] and the experimental data [8.16]

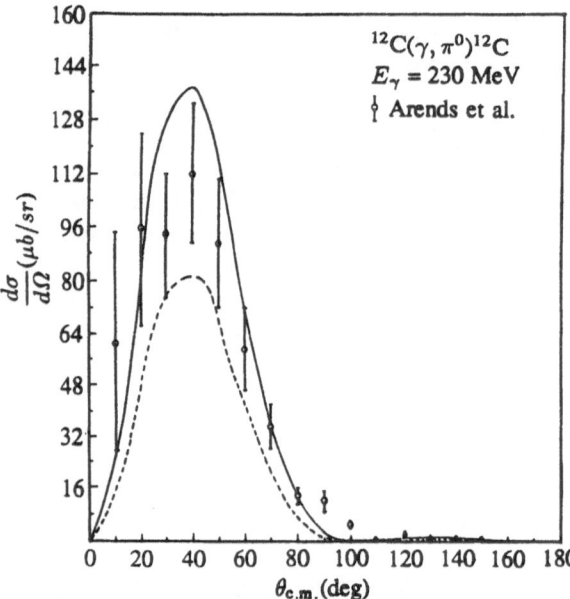

Fig. 8.2. Differential cross sections for $^{12}C(\gamma, \pi^0)\,^{12}C$ at $E_\gamma = 230\,\mathrm{MeV}$. DWIA calculations of *Girija* et al. [8.8] with CGLN (*solid line*) and BDW (*dashed line*) amplitudes. Experimental data from [8.16]

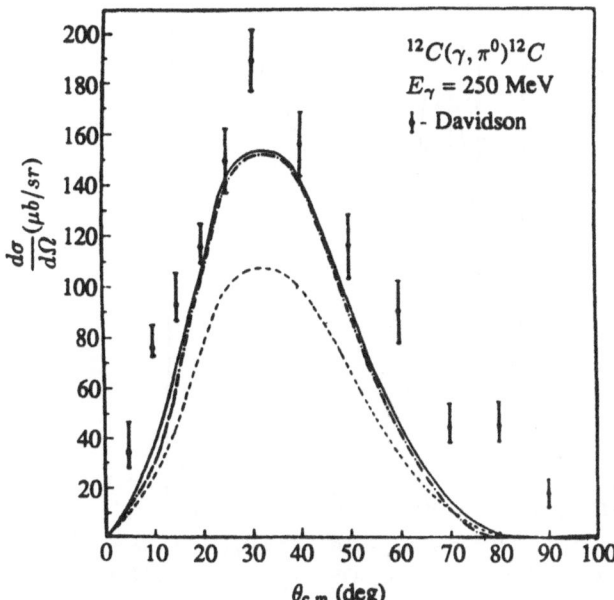

Fig. 8.3. Differential cross sections for $^{12}C(\gamma, \pi^0)\,^{12}C$ at $E_\gamma =$ 250 MeV. DWIA calculations of *Girija* et al. [8.8] with CGLN (*solid line*) and BDW (*dashed line*) along with IDM calculations (*dashed-dotted line*) of *Saharia* and *Woloshyn* [8.4]. Experimental data from Davidson [8.12]

4) The DWIA calculations made with the pion-nucleus optical potential of *Stricker* et al. [2.37] are in agreement with the experimental data and also with the theoretical calculation of *Saharia* and *Woloshin* [8.4] made with the isobardoorway model. Attention of the reader is drawn to Fig. 8.3.

Some of the above findings of GDNU [8.8] are supported by the subsequent DWIA calculations of *Kamalov* and *Kaipov* [8.9] and *Boffi* and *Mirando* [8.10].

Kamalov and Kaipov have done the DWIA calculations in momentum space which take into account the non-locality i.e., pion momentum dependence of the photoproduction operator. They have also included nucleon Fermi motion and medium modification of the Δ propagator. They have done calculations using both CGLN and BDW amplitudes and they find that the CGLN amplitude yields higher cross sections when compared to those obtained with the BDW amplitude at $E_\gamma = 230$ MeV and to a lesser extent at $E_\gamma = 290$ MeV. This is illustrated in Fig. 8.4 and this result is in support of the observation (3) although numerically the cross sections of Kamalov and Kaipov are higher. This may arise from the different prescriptions used in the construction of the nuclear amplitudes in the two calculations. The observation of Kamalov and Kaipov that the nucleon Fermi motion gives a reduction in the cross section is similar to the effect observed by *Koch* and *Moniz* [8.5,6] due to the medium modification in the $\Delta - h$ model.

Boffi and *Mirando* [8.10] have made DWIA calculations for ^{12}C, ^{16}O and ^{40}Ca at a lower energy range $E_\gamma \approx 135 - 180$ MeV, much below the Δ resonance and compared them with the experimental data from Saclay [8.17]. The DWIA calculation has been done in the coordinate space using the BL amplitude with Born terms, Δ terms and ω^0 exchange and replacing the pion momentum in the elementary amplitude by the gradient operator. The pion-nucleus optical potential

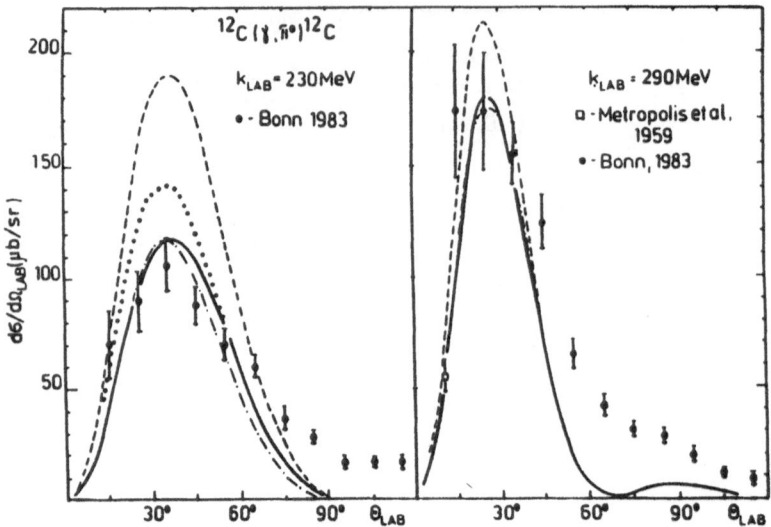

Fig. 8.4. Differential cross sections for $^{12}C(\gamma, \pi^0)\,^{12}C$ at $E_\gamma = 230$ and 290 MeV. Momentum space DWIA calculations of *Kamalov* and *Kaipov* [8.9]: *solid lines* (BDW amplitude); *dashed line* (CGLN amplitude); *dotted line* (CGLN amplitude, Fermi motion not included); *dash-dotted line* corresponds to $\Delta - h$ calculatioon [8.6,19]; experimental data from [8.16]

of SMC [2.37] has been used in GDNU [8.8]. Boffi and Mirando have made an important observation that in the energy region $E_\gamma = 135 - 180$ MeV, the cross sections obtained with DWIA are higher than the cross sections obtained with PWIA by about $50\% - 100\%$, depending on the incident γ-ray energy and also the optical potential used for the pion-nucleus interaction. They have illustrated this point for nuclei ^{12}C, ^{16}O and ^{40}Ca. This observation is consistent with the calculations of GDNU who have reported enhancement in the cross section due to distortion effects in the energy region below 230 MeV but reduction in the cross section for energies above 230 MeV.

8.3 The Isobar-Doorway Model

The isobar-doorway model of *Kisslinger* and *Wang* [8.1,2] rests on the observation that the pion-nucleon interaction in the medium energy range is dominated by the $\Delta(1232)$ baryon resonance in the $T = 3/2$, $J = 3/2$, $l = 1$ state. Restricting to s and p waves, the matrix element for the elementary $\pi - N$ scattering is of the form

$$\langle \boldsymbol{k}' | t(E) | \boldsymbol{k} \rangle_{\text{on-shell}} = a(E) + b(E)(\hat{\boldsymbol{k}}' \cdot \hat{\boldsymbol{k}}) \quad . \tag{8.17}$$

Since the p-wave is resonant,

$$b(E) \simeq \left[E - M_\Delta + \frac{i\Gamma_\Delta(E)}{2} \right]^{-1} \quad , \tag{8.18}$$

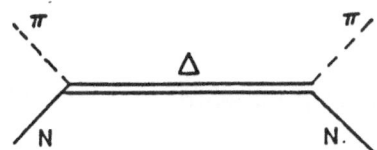

Fig. 8.5. π-nucleon t matrix in the isobar model with the formation of a Δ-isobar in the intermediate state

where M_Δ and Γ_Δ denote the isobar mass and width. The matrix element for $\pi - N$ scattering can now be conveniently written as a sum of two parts, one non-resonant (t^{NR}) and the other resonant (t^R)

$$\langle k'|t|k \rangle = t^{NR}(k', k, E) + t^R(k', k, E) \quad . \tag{8.19}$$

The resonant part is the dominant one and it proceeds through the formation of Δ isobar in the intermediate state as shown in Fig. 8.5.

The resonant part of the t-matrix of $\pi - N$ scattering can be written as

$$t^R(k', k, E) = \frac{\langle \pi(k') N(P')|H|\Delta \rangle \langle \Delta |H|\pi(k) N(P) \rangle}{E - M_\Delta + \Gamma_\Delta(E)/2} \quad . \tag{8.20}$$

The pion-nucleus T-matrix in the isobar doorway model is written analogously to the pion-nucleon t-matrix given in (8.19,20).

$$T = T_{NR} + \sum_i \frac{\langle \psi_f|H|D_i \rangle \langle D_i|H_i|\psi_0 \rangle}{E - E_{D_i} + i\,\Gamma_i(E)/2} \quad , \tag{8.21}$$

where D_i denotes the isobar-doorway states consisting of a Δ and various number of particles and holes. These states are quite complicated but the unknown dynamics is represented by a set of parameters.

The numerator in the second term of (8.21) can be written in terms of a single nucleon t-matrix and a nuclear-isobar form factor using the impulse approximation and the shell-model description for the nucleus.

$$\begin{aligned} N_i &= \langle \psi_f|H|D_i \rangle \langle D_i|H|\psi_0 \rangle \\ &= \sum_N \int d^3p\, d^3p'\, \phi_N^*(p')\, \phi_i(p' + k')\, \phi_i^*(p + k)\, \phi_N(p) \\ &\quad \times \langle p'k'|H|\Delta \rangle \langle \Delta|H|pk \rangle \end{aligned} \tag{8.22}$$

where ϕ_N denotes the single nucleon orbital state in the momentum space and ϕ_i, the wave function of the Δ in the doorway state D_i. From (8.20), it follows

$$\langle p'k'|H|\Delta \rangle \langle \Delta|H|pk \rangle = \left(E - M_\Delta + \tfrac{1}{2}\,\Gamma_\Delta(E) \right) \langle k'|t^R(E)|k \rangle \quad . \tag{8.23}$$

Substituting (8.23) into (8.22), we get

$$N_i = \left(E - M_\Delta + \tfrac{1}{2}\,\Gamma_\Delta(E) \right) \langle k'|t^R(E)|k \rangle\, F_i(k', k) \quad , \tag{8.24}$$

117

where $F_i(k', k)$ is the nuclear-isobar form factor

$$F_i(k', k) = \sum_N \int d^3p\, d^3p'\, \phi_N^*(p')\, \phi_i(p' + k')\, \phi_i^*(p + k)\, \phi_N(p) \ . \tag{8.25}$$

Using the result (8.24), (8.21) is rewritten as

$$\begin{aligned}
\langle k'|T|k\rangle &= \langle k'|T_{NR}|k\rangle \\
&+ \sum_i \frac{(E - M_\Delta + i\Gamma_\Delta/2)\, \langle k'|t^R(E)|k\rangle}{E - E_{D_i} + i\Gamma_i(E)/2}\, F_i(k', k) \ .
\end{aligned} \tag{8.26}$$

This is the main result of *Kisslinger* and *Wang* [8.2] in the formulation of the isobar-doorway model. The determination of the quantities E_{D_i}, $\Gamma_i(E)$ and $F_i(k', k)$ requires a detailed understanding of the isobar dynamics involving the construction of doorway states and their coupling to the complicated many body states. But in the phenomenological approach of the isobar-doorway model, these are treated as parameters chosen to yield the pion-nucleus elastic cross section. It is assumed that the width Γ_i of the doorway states is larger than the average separation energy so that the energy denominator in (8.26) can be averaged over the contributing states and also the nuclear form factor is modified taking into account the non-locality associated with the Δ propagation in the nucleus.

Now the problem reduces to the evaluation of the integral in (8.25) and the determination of the nuclear-isobar form factor. The momentum conservation $p + k = p' + k'$ makes the integration over d^3p' redundant. Writing all the momentum space wave functions in (8.25) in configuration space, the integral over d^3p yields the delta function $\delta(r_1 - r_2 - r_3 + r_4)$, thereby leaving only two of the variables r_1 and r_2 independent. Now

$$\begin{aligned}
F_i(k', k) &= \sum_N d^3r_1\, d^3r_2\, e^{ik' \cdot r_1}\, e^{-ik \cdot r_2} \\
&\times \phi_N^*(r_1)\, \phi_N(r_2)\, \phi_i(r_1)\, \phi_i^*(r_2) \quad .
\end{aligned} \tag{8.27}$$

The isobar density function

$$\varrho_\Delta(r_1, r_2) = \sum_i \phi_i(r_1)\, \phi_i^*(r_2) \tag{8.28}$$

is non-local. Many convenient forms have been tried for $\varrho_\Delta(r_1, r_2)$. *Kisslinger* and *Wang* [8.2] have used the following choice of the form factor for the numerical computation neglecting non-locality

$$F(k', k) = \sum_l i^l(2l + 1)^2 \int dr\, r^2\, j_l(qr)\, \varrho_N(r)\, P_l(\cos \theta) \tag{8.29}$$

where $q = k' - k$.

Saharia and *Woloshin* [8.4] have chosen the following isobar density matrix

$$\varrho_\Delta(r_1, r_2) = \sum_i \phi_i(r_1)\, \phi_i^*(r_2)$$

$$= \exp\left[-(r_1 - r_2)^2/\lambda^2\right]\Big/(\pi\lambda^2)^{3/2} \tag{8.30}$$

where λ is a parameter which defines non-locality

$$\varrho_\Delta(r_1, r_2) \to \delta^3(r_1 - r_2) \quad , \qquad \text{as} \quad \lambda \to 0 \quad . \tag{8.31}$$

Using this isobar density, the form factor $F(k', k)$ has been evaluated for $1p$ shell nuclei using harmonic oscillator wave functions. Besides the non-locality paramter λ, let us introduce two more parameters ΔE, the energy shift and β, the width parameter and rewrite (8.26)

$$\langle k'|T|k \rangle = \langle k'|T_{\text{NR}}|k \rangle$$

$$+ \frac{(E - M_\Delta + i\Gamma_\Delta/2)\,\langle k'|t^R(E)|k \rangle}{E - M_\Delta - \Delta E + i\beta\Gamma_\Delta(E)/2}\, F(\lambda, k', k) \quad . \tag{8.32}$$

The parameters ΔE, β and λ of the isobar-doorway model take into account not only the various many-body effects like true absorption, Pauli blocking and inelastic scattering but also the effect of Fermi motion. The parameters chosen by Saharia and Woloshyn to fit the pion-nucleus elastic scattering data were tabulated in [8.4].

Hitherto, we have discussed the isobar-doorway formation for the pion-nucleus scattering and the extension to the coherent π^0 photoproduction in nuclei is straightforward. The transition matrix for (γ, π^0) reaction in nuclei can be written down following the same procedure used in the construction of the T-matrix (8.32) for the pion-nucleus elastic scattering

$$\langle q|T|k \rangle = t^{\text{Nr}}_{\gamma\pi^0}(q, k, E)\, \varrho(q - k)$$

$$+ \frac{E - M_\Delta + i\Gamma_\Delta(E)/2}{E - M_\Delta - \Delta E + i\beta\Gamma_\Delta/2}\, t^R_{\gamma\pi^0}(q, k, E)\, F(\lambda, q, k) \;, \tag{8.33}$$

where the quantities ΔE and β are the same as those that occur in the T-matrix of pion-nucleus elastic scattering and also the form of the nuclear-isobar form factor F is the same. This is because in coherent π^0 photoproduction, the same doorway states occur as in elastic scattering and the quantities ΔE, β and F define the properties of these doorway states. The only additional inputs are the appropriate single nucleon amplitudes.

Strictly speaking, the T-matrix defined above is correct only to first order and it has to be iterated to all orders to yield the pion-nucleus scattering T-matrix (denoted by $\tilde{T}_{\pi\pi}$) and the T-matrix for the pion photoproduction reaction (denoted by $\tilde{T}_{\gamma\pi^0}$)

$$\tilde{T}_{\pi\pi} = \left(1 + \tilde{T}_{\pi\pi}\, G_0\right) T_{\pi\pi} \tag{8.34}$$

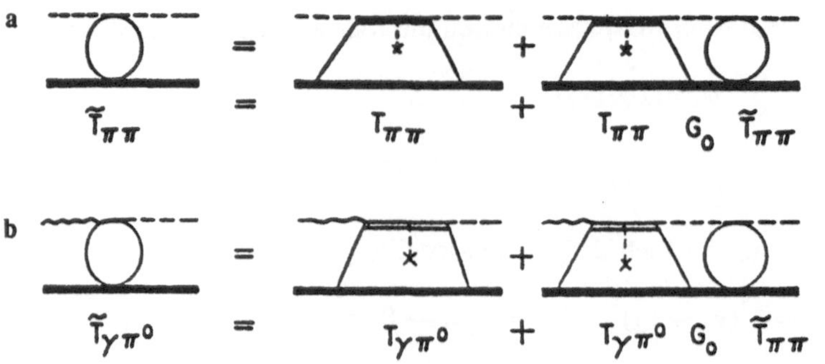

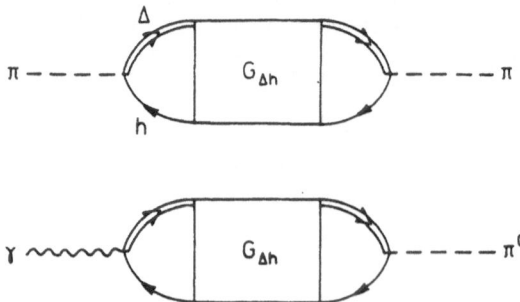

Fig. 8.6. Diagrammatic representation of pion elastic scattering and π^0 photoproduction in the isobar-doorway model. The interaction $---\times$ denotes all many-body modifications to the isobar propagator

$$\tilde{T}_{\gamma \pi^0} = \left(1 + \tilde{T}_{\pi\pi} G_0\right) T_{\gamma \pi^0} \tag{8.35}$$

where $G_0 = \left(E - H_0\right)^{-1}$ is the free Green's function. This is illustrated in Fig. 8.6.

8.4 The Isobar-Hole Formalism

Just as the isobar-doorway model, the isobar-hole formalism assumes that the pion induced or gamma induced reactions in the (3,3) resonance region predominantly go through the excitation of the nucleus from its ground state into the isobar-hole states. This is illustrated in Fig. 8.7. The Δ-hole propagator is modified by an interaction with the medium and $G_{\Delta b}$ denotes the 'dressed' propagator.

First let us reformulate the multiple scattering theory of pion-nucleus elastic scattering so as to form a basis for both the isobar-doorway model and the isobar-hole formalism. For this, we need to introduce the projection operators P, D and Q. The operator P projects the space consisting of pion and the ground state nucleus, D the space consisting of isobar-hole states and Q all the rest not included in either P or D. Given the doorway condition $H_{PQ} = 0$, the pion-nucleus elastic transition matrix is given in [8.5,6]

Fig. 8.7. Diagrammatic representation of (a) pion-nucleus scattering and (b) photoproduction of pions from a nucleus

120

$$T_{PP} = H_{PD} \, G_{\Delta h} \, H_{DP}$$
$$= H_{PD} \left[E - E_R + i\Gamma(E)/2 - \mathcal{H}_{\Delta h} \right]^{-1} H_{DP}$$

$$\mathcal{H}_{\Delta h} = H_{DD} + H^{\uparrow}_{DD} + H^{\downarrow}_{DD} \tag{8.36}$$

$$H^{\uparrow}_{DD} = H_{DP} \left[E^+ - H_{PP} \right]^{-1} H_{PD}$$

$$H^{\downarrow}_{DD} = H_{DQ} \left[E^+ - H_{QQ} \right]^{-1} H_{QD} \quad .$$

The isobar-hole Hamiltonian consists of three parts. The diagonal interaction H_{DD} includes isobar propagation, binding effects, Pauli blocking effects and isobar-hole residual interaction. The interaction $H^{\uparrow}_{DD}$ generates the elastic width and $H^{\downarrow}_{DD}$ the spreading width due to the intermediate coupling to the more complicated Q space.

The normalized doorway state for the pion-nucleus partial wave is defined as

$$\left| D^L_0 \right\rangle = N_L \, H_{DP} \left| (q)_L ; 0 \right\rangle \quad , \tag{8.37}$$

where $\left| (q)_L ; 0 \right\rangle$ represents the partial wave projection of the pion plane wave and the nuclear ground state. It follows that

$$N_L = \left\langle (q)_L ; 0 \left| H_{PD} \, H_{DP} \right| (q)_L ; 0 \right\rangle^{-1/2} \quad . \tag{8.38}$$

The doorway state is a linear superposition of $\Delta - h$ states. The pion-nucleus partial wave transition matrix is simply the expectation value of $G_{\Delta h}$ in the doorway state

$$T_L = N^{-2}_L \left\langle D^L_0 \left| G_{\Delta h} \right| D^L_0 \right\rangle \quad . \tag{8.39}$$

The isobar-doorway model of *Kisslinger* and *Wang* [8.1,2] assumes that the doorway state is an eigenstate of the Hamiltonian $\mathcal{H}_{\Delta h}$

$$T^{\text{Doorway}}_L = \frac{N^{-2}_L}{E - E_R + i\Gamma/2 - \left\langle D^L_0 \left| \mathcal{H}_{\Delta h} \right| D^L_0 \right\rangle} \quad . \tag{8.40}$$

In the isobar-hole formalism of *Koch* and *Moniz* [8.5,6], the doorway state is not an eigenstate of the Hamiltonian $\mathcal{H}_{\Delta h}$ but is constructed by repeated application of the Hamiltonian $\mathcal{H}_{\Delta h}$ starting from D^L_0

$$\left| D^L_i \right\rangle = N^L_i \left| \mathcal{H}_{\Delta h} \right| D^L_{i-1} \right\rangle - \sum_{j=0}^{n-1} \left| D^L_{i-1} \right\rangle \left\langle D^L_{i-1} \left| \mathcal{H}_{\Delta h} \right| D^L_j \right\rangle \tag{8.41}$$

where N^L_i is a normalization factor. The resulting transition matrix is

$$T_L = \cfrac{N^{-2}_L}{E - E_R + i\Gamma/2 - \mathcal{H}^L_{00} - \cfrac{\left(\mathcal{H}^L_{01} \right)^2}{E - E_R + i\Gamma/2 - \mathcal{H}^L_{11} - \cfrac{\left(\mathcal{H}^L_{12} \right)^2}{E - E_R + i\Gamma/2 - \mathcal{H}^L_{22} - \cdots}}} \quad , \tag{8.42}$$

where

$$\mathcal{H}_{ij}^L = \langle D_i^L | \mathcal{H}_{\Delta h} | D_j^L \rangle \quad . \tag{8.43}$$

The first term in the continued fraction of (8.42) conforms to the isobar-doorway model. It has been shown by *Koch* and *Moniz* [8.5,6] that the continued fraction has rapid convergence properties for elastic scattering and the inclusion of only 3 doorway states gives an accuracy of 1 %.

The $\Delta - h$ formalism outlined above for pion-nucleus elastic scattering can be directly extended to the study of coherent π^0 photoproduction [8.5,6] by writing the transition matrix in an analogous way

$$T_{\gamma\pi^0} = H_{PD} \frac{1}{E - E_R + i\Gamma/2 - \mathcal{H}_{\Delta h}} H_{D\gamma} \quad . \tag{8.44}$$

Only we need to know the vertex functions coupling the pion to the $\Delta - h$ state and the photon to the $\Delta - h$ state

$$H_{DP} |(q)_L ; 0\rangle = N_L F_{\pi N\Delta} |(q)_L ; 0\rangle \tag{8.45}$$

$$H_{D\gamma} |(k\lambda)_L ; 0\rangle = N_L^\gamma F_{\gamma N\Delta} |(k\lambda)_L ; 0\rangle \quad . \tag{8.46}$$

The vertex couplings $F_{\pi N\Delta}$ and $F_{\gamma N\Delta}$ are given by

$$F_{\pi N\Delta}^+ = \frac{g_{\pi N\Delta}}{M_\Delta} \boldsymbol{q} \cdot \boldsymbol{S} \, T_\alpha \, v(q^2) \tag{8.47}$$

with

$$v(q^2) = \left(1 + \frac{q^2}{\alpha^2}\right)^{-1} \tag{8.48}$$

and

$$F_{\gamma N\Delta} = \frac{g_{\gamma N\Delta}}{M_\Delta} \boldsymbol{\varepsilon}_{k\lambda} \times \boldsymbol{k} \cdot \boldsymbol{S}^+ \, T_3^+ \quad . \tag{8.49}$$

The spin transition operator S transforms the nucleon (spin 1/2) into an isobar (spin 3/2) and it obeys the relation

$$S_i^+ S_j = \tfrac{2}{3} \delta_{ij} - \tfrac{i}{3} \varepsilon_{ijk}\sigma_k \quad . \tag{8.50}$$

The isospin operator is denoted by T.

Koch and *Moniz* [8.5,6] have considered the coherent π^0 photoproduction from ^{12}C, ^{16}O and ^{4}He and obtained reasonable agreement with the available experimental data. *Takaki, Suziki* and *Koch* [8.7] have extended the investigation to include incoherent π^0 photoproduction from ^{12}C leading to low-lying excited states since the Bonn experiment is contaminated by such incoherent contributions due to finite energy resolution of the π^0 spectrometer. At 295 MeV, the coherent result (solid curve in Fig. 8.8a) is clearly too low to explain the data. By adding the contributions of the transitions to the 2^+ (4.44 MeV) state (dashed line), the 3^- (9.64 MeV) state (dash-dotted line) and the 0^+ (7.65 MeV) state (dotted line), the solid line in Fig. 8.8b is obtained, leading to significantly improved agreement with the data.

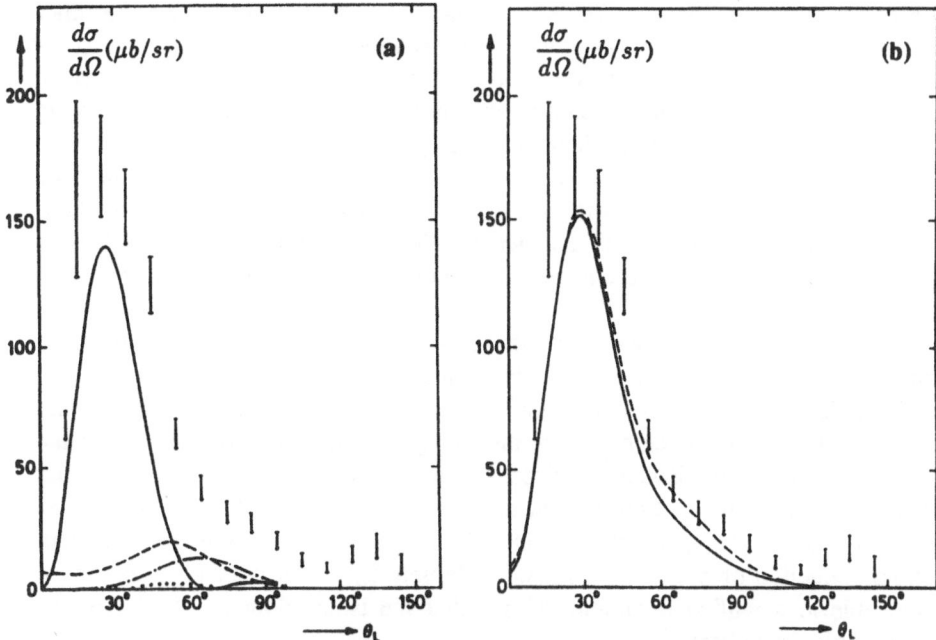

Fig. 8.8. Coherent and incoherent pion photoproduction on ^{12}C at $E_\gamma = 295$ MeV. Results of $\Delta - h$ calculations [8.7] along with the experimental data [8.16]. (a) Contributions from individual final states. (b) Summed cross section (*solid line*)

Takaki et al.[8.7] have also made a detailed study of the medium modification of the production amplitude, the effects of Pauli blocking on Δ-decay inside the nucleus, the Δ recoil and binding corrections in the propagator and the net effect of all these is to suppress the nuclear π^0 photoproduction for the nuclear transitions, they considered. For further details, the reader is referred to [8.5–7].

8.5 Concluding Remarks

We reiterate that the coherent π^0 photoproduction from spin zero nuclei holds out great promise in precisely determining the spin-independent part of the elementary amplitude besides yielding valuable information on the production and propagation of isobars in a nuclear medium. The isobar doorway model of *Kisslinger* and *Wang* [8.1,2] and the isobar-hole formalism of *Koch* and *Moniz* [8.5,6] are essentially the same although they differ in details and they have been applied with great success to a wide variety of processes involving photons and pions in the intermediate energy range [8.19]. The DWIA formalism reported in Sect. 8.2 is equally successful in explaining the rich variety of reaction processes by choosing properly the optical potential which takes into account the first and second order corrections. Since the DWIA calculations of GDNU [8.8] and the other calculations using the isobar doorway model and the isobar-hole formalism compare with each other numerically, it appears that all these calculations include the medium effects on the production operator and the final state interactions to the same extent. It is also aesthetically satisfying that the DWIA method which has been successful to a large extent in explaining the charged photo-pion cross sections is equally successful in predicting the coherent π^0 cross sections. Otherwise, one will be faced with a larger question — why the DWIA which is so successful in explaining the charged photopion data fails in the case of coherent π^0 photoproduction?

The isobar doorway model and the $\Delta - h$ model have a distinct advantage since these models provide a clear visual picture of the production, propagation and the decay of the Δ isobar within the nucleus. But these models are greatly restricted to an energy region where the background contributions are negligible when compared to the isobar contribution and to the processes in which the isobar contribution dominates. Fortunately, there is a wide variety of reactions such as Compton scattering, photoabsorption, elastic scattering of pions and coherent photoproduction of π^0, that are mediated largely by isobar doorways and hence amenable to a unified treatment using either the isobar-doorway model or the isobar-hole formalism.

Let us conclude this section with a few remarks. The DWIA method is universal and is applicable to a wide variety of phenomena wherein strong interactions are involved. Improvements have been incorporated continuously by including higher order corrections in the optical potential and medium modifications of the transition operator. The isobar-doorway model and the isobar-hole formalism are applicable only to selected processes wherein the isobar channel is the dominant one. Just as the Bohr model of the atom offers a better visual picture of how the spectral lines are emitted than the quantum mechanical treatment, the isobar doorway hypothesis holds out a greater appeal to those who wish to visualise the reaction process. There is bound to be an increasing application of the isobar-doorway hypothesis to a rich variety of phenomena in the intermediate energy range and it is bound to develop into a powerful tool in the analysis of the forthcoming experiments in this energy range [8.20]. In the recent workshop on 'photon and neutral meson physics at intermediate energies' held at Los Alamos during January $7-9$, 1987, a great emphasis was laid on the study of neutral meson photoproduction and it advocated strongly the building of a greatly improved π^0 spectrometer [8.21]. Experiments on π^0 photoproduction are in progress in Bates, Bonn, Mainz, Saclay and Saskatchewan and with the prospects of new experimental data with improved precision coming in, the theoretical study should receive the necessary stimulus to resolve some of the puzzles that exist at present. What criteria should one apply to choose between the several amplitudes, all of which yield correctly the cross section for the elementary process but differ in the yield of coherent π^0 photoproduction cross section from closed shell nuclei? Since different groups work with different elementary amplitudes and use either the DWIA formalism with a wide variety of the pion-nucleus optical potential or the isobar doorway formalism, it is hard to extract any unambiguous information regarding the medium modification of the transition operator on the effect of distortion on the outgoing pion. Should one include both these effects? Does some of the second-order pion-nucleus optical potentials include medium corrections and if so, is there a possibility of double counting?

9. Special Topics and Future Prospects

We shall discuss, in this concluding chapter, special topics such as polarization phenomena in photopion nuclear physics, electroproduction of pions and photoproduction of more exotic mesons such as K and η. Polarization studies throw light on the detailed structure of photopion production amplitude and the study of electroproduction of pions on the extent of contribution of the longitudinal virtual photons. The photoproduction of K mesons leads to the production of hypernuclei and the possibility of extracting information on the $\Delta - N$ interaction whereas the photoproduction of η mesons is purely of academic interest, at present. We conclude this review by assessing the progress made so far in the study of photoproduction of mesons with the existing facilities and indicating the future prospects with the advent of 100 % duty cycle electron machines at higher energies, improved detector techniques and high resolution pion spectrometers.

9.1 Polarization Studies

Any reaction amplitude on a spin$-1/2$ system will have a gross structure

$$t = \boldsymbol{\sigma} \cdot \boldsymbol{K} + L \tag{9.1}$$

and the elementary amplitude for photoproduction of pions on single nucleon can also be written in that form. The differential cross sections for the process will be proportional to the sum of the squares of the spin-dependent and spin-independent amplitudes.

$$\frac{d\sigma}{d\Omega} \approx |K|^2 + |L|^2 \quad . \tag{9.2}$$

But the polarization P of the recoil nucleon will involve cross terms since one has to construct a pseudovector quantity

$$P = \frac{\mathrm{Tr}(\sigma \varrho)}{\mathrm{Tr}\,\varrho} = \frac{\mathrm{i}(\boldsymbol{K} \times \boldsymbol{K}^* + L^*\boldsymbol{K} + L\boldsymbol{K}^*)}{\boldsymbol{K} \cdot \boldsymbol{K}^* + LL^*} \quad . \tag{9.3}$$

So, by conducting the polarization study on the recoiling nucleon, one is able to test a different bilinear combination of the amplitudes K and L and thus able to obtain a greater insight into the structure of such amplitudes. The recoil nucleon is polarized even when the incident photon is not polarized.

In a similar way, one can obtain a more detailed information on the structure of the amplitude using polarized photons. *Bernstein* [1.10] has suggested how to obtain information on the $E0$ and $M1$ terms by using polarized photons. Writing the transition amplitude in terms of the polarization vector ε of the photon

$$t = a\,\varepsilon \cdot (k \times q) + ib\,\sigma \cdot \varepsilon + i\varepsilon \cdot q\,[c(\sigma \cdot k) + d(\sigma \cdot q)] \quad , \tag{9.4}$$

we obtain an expression for the cross section as given below

$$\sigma(\theta, \phi) \propto |a|^2 \sin^2 \theta \cos^2 \phi + |b|^2 + \Delta(\theta) \sin^2 \phi \quad ,$$

when θ is the direction of the emitted pion and ϕ, the angle between ε and the normal n to the scattering plane ($n = (k \times q)/|k \times q|$). We see that the contribution from the Kroll–Ruderman term $|b|^2$ is independent of ϕ and that the $E0$ term $|a|^2$ vanishes at $\phi = \pi/2$ (i.e. if ε lies in the scattering plane) while the momentum dependent $M1$ term $\Delta(\theta)$ vanishes at $\phi = 0$. Thus by making measurement at three angles of ϕ, we can determine all the three components.

In the same way, we can obtain further information by using unpolarized photon on polarized targets. If the nucleon polarization σ is chosen perpendicular to k or q, then the momentum dependent terms c and d are eliminated. By choosing σ in between k and q, the relative sign can be fixed. With both polarized photons and polarized targets, the momentum independent term can be eliminated by orienting σ perpendicular to ε.

Thus, by conducting the polarization study, we are able to check the finer details of the transition amplitude and assess the relative strengths of the various terms that contribute to the process.

Ramachandran and *Devanathan* [9.1,2] more than two decades ago suggested such a polarization study in the pion photoproduction process with complex nuclei. It has been shown by them [9.1] that in the case of nuclear transition to bound states the final residual nucleus is polarized even if the incident photon is unpolarized. The recoil nuclear polarization is perpendicular to the plane containing the vectors k and q. It is not surprising since the only axial vector that is available for the problem under consideration is $(k \times q)$. *Ramachandran* and *Devanathan* [9.1] considered a specific reaction

$$\gamma + {}^{29}\mathrm{Si}_{14} \rightarrow \pi^- + {}^{29}\mathrm{P}_{15} \tag{9.5}$$

for illustrative purpose and they have made an interesting observation that the recoil nuclear polarization as a function of the emission angle of the pion does not depend upon the nuclear wave function. They have used an independent particle model for the nucleus and even if a more realistic configuration mixing model is used, the result will be only weakly dependent on nuclear structure. At this juncture, it will be fruitful to recall a similar situation that has been pointed out by *Devanathan* et al. [9.3] in the context of muon capture and the experimental study that has been soon undertaken by the Louvain – Saclay – ETH collaboration [9.4] and later by the Tokyo group [9.5] to determine the muon capture interaction coupling constants without the uncertainties of nuclear structure.

In a nuclear problem, if it is possible to find an observable that is to a large extent independent of nuclear structure, then that observable can be used to investigate the medium modification of the elementary transition amplitude. *Ramachandran* and *Devanathan* [9.1] used the CGLN amplitude for the photopion production and this amplitude involves the magnetic moments of the proton and neutron as input. It is known that the magnetic moment of the bound nucleon in a nucleus is to some extent quenched and this quenching may be considered as due to medium effects. This has been investigated and the recoil nuclear polarization is found to be sensitive to the quenching of the magnetic moment.

It is desirable to reinvestigate the recoil nuclear polarization in depth and look into the effects of final state interaction, nuclear structure and medium modifications of the elementary amplitude. The polarization study of *Ramachandran* and *Devanathan* [9.1] involves the detection of the outgoing pion and the measurements of the polarization of the recoil nucleus simultaneously. This is a difficult task. But it will definitely be a rewarding exercise to work out a simple geometry at which the polarization experiment will be feasible. The preliminary theoretical studies are already encouraging.

9.2 Electroproduction of Pions

Most of the theoretical calculations have been made on (γ, π) reactions whereas the experimental studies have been conducted on pion electroproduction (e, π). In recent experiments [9.6–14], the double differential cross sections $d^2\sigma/dT_\pi d\Omega_\pi$ have been measured with incident electrons of specified energy without detecting the outgoing electrons. The experimental data have been analyzed using virtual photon theory in which the electromagnetic interaction of the incident electron with the target nucleus (N) is represented by the interaction of the virtual photon (Fig. 9.1). From the experimental data, the (γ, π) differential cross sections have been deduced using the virtual photon spectrum

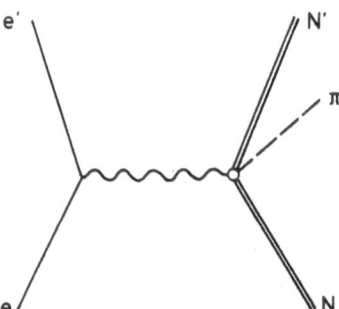

Fig. 9.1. Electroproduction of pion from nucleus

$$\left(\frac{d^2\sigma}{d\Omega_\pi \, dT_\pi}\right)_{e,\pi} = \sum_{E_f} \left(\frac{d\sigma(k,E_f)}{d\Omega}\right)_{\gamma,\pi} N_{h\nu}(k,T_e) \quad , \tag{9.6}$$

where T and T_e denote the kinetic energy of the produced pion and incident electron respectively, E_f is the energy of the residual states f, $N_{h\nu}(k,T_e)$ is the virtual photon spectrum associated with the incident electron. Usually the virtual photon spectrum is calculated using the formula of *Dalitz* and *Yennie* [9.15] but an improvement to this procedure has been suggested recently by *Stoler* et al.[9.16], *Tiator* and *Drechsel* [9.17], *Tiator* and *Wright* [9.18] and others [9.19,20].

On the other hand, a direct theoretical investigation of (e, π) reactions and the calculation of the cross section $d^2\sigma/dT_\pi \, d\Omega_\pi$ using the full elementary amplitude for electroproduction of pions is desirable. Such a program has been recently undertaken by *Sabutis* and *Tabakin* [9.21] who have made an exhaustive study of electroproduction of pions from light nuclei. They have deduced an amplitude for pion photoproduction by virtual photons using the Born diagrams and adding a Δ contribution, following closely the formalism of *Blomqvist* and *Laget* [2.45]. The pion electroproduction amplitude will be useful in assessing the contribution of longitudinal virtual photons in double arm experiments wherein the scattered electron and the emitted pion are detected in coincidence (Fig. 9.1). Such experiments do not exist at present and so *Sabutis* and *Tabakin* [9.21] have made the calculation for a single arm experiment wherein the emitted pion alone is detected. They have confirmed that in such an experiment, contributions to the cross sections from longitudinal virtual photon components are negligible.

On the other hand to assess the contribution from the longitudinal virtual photon, it will be more fruitful to design single arm experiments wherein the outgoing electron is detected and a right guess is made of the final nuclear state when the energy transfer is greater than the pion production threshold. By this, we fix the four-momentum transfer from the electron and hence it is possible to learn about the contribution of the longitudinal photons. Such calculations have been made earlier by *Devanathan* [9.22,23] using the electroproduction amplitude of *Fubini* et al. [9.24]. The deuteron is a simple target for which pion electroproduction calculations have been made by him [9.22], using the closure approximation for the two-nucleon final state and the comparison made with the experimental results of *Ohlsen* [9.25]. Several other claculations on electroproduction of pions have also been made. The earliest one was that of *Czyz* and *Walecka* [9.26] using the Fermi gas model for the nucleus. *Borie* et al. [9.27] used a configuration mixing model for the nucleus and the final state interactions for the pion with the residual nucleus. More recent study includes those of *Borkowski* et al. [9.28], *Dressler* et al. [9.29] and *Cohen* and *Eisenberg* [9.30]. *Asai* et al. [9.31], *Tiator* and *Drechsel* [9.17], *Skopic* et al. [9.32] have investigated the electroproduction of pions on the three-nucleon system.

More work will be done in the near future. *Bernstein* [1.10] has referred to the prospect of using internal targets both at Bates and NIKHEF that will permit the observation of electroproduction of very low energy pions by the detection of the recoil product nucleus from an ultra thin target. The reaction ^{3}He$(e,e',\pi)^3$H is very favorable for study since the cross section is found to be sufficiently large [1.10] for a measurement with internal target facilities under construction.

With the new CW machines that are still under construction, there are better chances of performing the double arm coincidence experiment and then only it will be possible to test the theory of electroproduction of pions fully and assess the contribution that comes from the longitudinal component of the virtual photon.

9.3 Photoproduction of K and η Mesons

Hitherto, we have discussed the photo- and electroproduction of pions. Now we shall make a brief reference to the photoproduction of more exotic mesons.

At higher photon energies, K^+ meson (mass = 493.67 MeV, strangeness = +1) can be produced

$$\gamma + p \rightarrow K^+ + \Lambda^0 \quad .$$

The proton is converted into Λ^0 particle (mass = 1115.6 MeV, strangeness = -1). This is known as associated production of strange particles. The amplitude for photoproduction of K mesons can be written in the same form as the CGLN amplitude for photopion production since the K^+ meson like the π^+ meson is a spinless particle with intrinsic odd parity. It is an exciting possibility to produce hypernuclei by the photoproduction of K^+ mesons and populate all possible hypernuclear states by spin-flip transition. Recently, several studies have been made on this exciting possibility and *Rosenthal* et al. [9.33,34] have completely investigated the final state interaction and calculated the cross section for production of hypernuclei in different states. So, photoproduction of K mesons leads to an interesting study of the optical potential of Kaon-nucleus interaction and also to the investigation of the structure of the hypernucleus. Besides Λ^0 hypernuclei, Σ^0 hypernuclei can also be produced in the photoproduction of K mesons since the reaction

$$\gamma + p \rightarrow K^+ + \Sigma^0$$

is equally possible.

Lastly, we shall turn to the photoproduction of η mesons

$$\gamma + p \rightarrow \eta + p \quad .$$

The η meson has a mass of 548.8 MeV and is also a member of the family of octet mesons (Fig. 9.2). Since the η meson is an isoscalar, it can only produce a baryon resonance N* with isospin $T = 1/2$. Experiments have been performed studying η production at LAMPF with pions [9.35] and protons [9.36,37], at Saclay with protons [9.38] and at LEAR with antiprotons [9.39]. Photoproduction of η mesons on a proton target is scheduled at Bates shortly and many more experiments of this nature may be performed at CEBAF when it goes into operation.

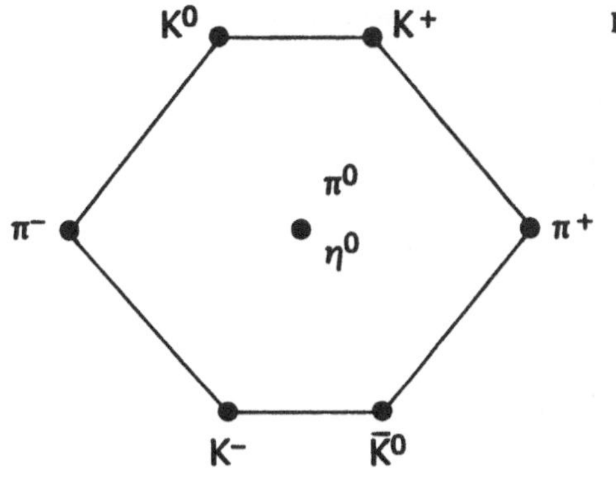

Fig. 9.2. Octet mesons

The amplitude for photoproduction of K and η mesons has the same structure as that of pion photoproduction and a form similar to that of CGLN can be used for investigation

$$t = F_1 \boldsymbol{\sigma} \cdot \boldsymbol{\varepsilon} + F_2 i \boldsymbol{\sigma} \cdot \boldsymbol{q} \, \boldsymbol{\sigma} \cdot \boldsymbol{\varepsilon} \times \boldsymbol{k} + F_3 \boldsymbol{\sigma} \cdot \boldsymbol{k} \, \boldsymbol{q} \cdot \boldsymbol{\varepsilon} + F_4 \boldsymbol{\sigma} \cdot \boldsymbol{q} \, \boldsymbol{q} \cdot \boldsymbol{\varepsilon} \quad .$$

Using a similar amplitude, *Tabakin* et al. [9.40] have investigated the photoproduction of η mesons from complex nuclei and the reader is referred to that article [9.40] and that of *Hicks* [9.41] for further details.

9.4 The Present Status and Future Prospects

The existing scenario with regard to electron accelerators is clearly depicted in Fig. 9.3. The major facility that will be added in the near future is the Continuous Energy Beam Accelerator Facility (CEBAF) and its unique place in the energy versus duty factor domain is clearly illustrated [9.42−44].

The CEBAF is being designated to fulfill the following specifications to meet the varying physics requirements:

1) Electron energy variable from 0.5 to 4.0 GeV
2) 100 % duty factor
3) Beam intensity of 200 μA
4) Three simultaneous beams with correlated energies.

It is expected to be commissioned in the year 1993.

Already the existing electron machines have yielded fruitful results in spite of the limitations. Most of the angular distribution measurements for $(\gamma, \pi^{\pm})$ reactions have come from Tohoku Linac at $E_\gamma \approx 200$ MeV. These experiments have provided a greater insight into the mechanism of photoproduction of charged pions from nuclei besides shedding light on the nuclear structure and the pion-nucleus optical potential. Due to the energy limitation of this machine, it cannot be used to study photopion reaction at higher gamma ray energy.

131

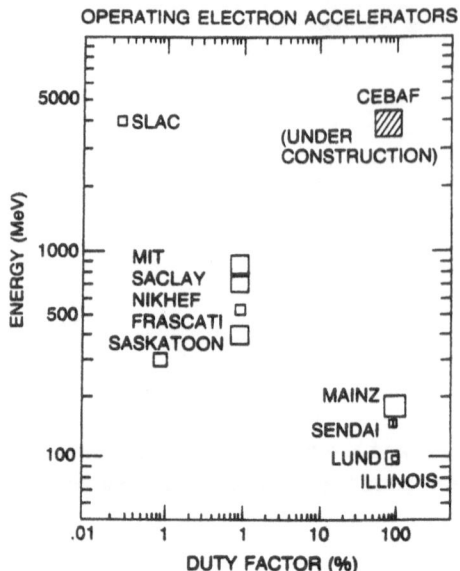

OPERATING ELECTRON ACCELERATORS

Fig. 9.3. Scenario of existing and proposed electron accelerators

Similarly, some useful information has been obtained at the pion threshold region from the CW machine (100 % duty cycle) working at Mainz for several years at 180 MeV. The (γ, π^0) cross section from proton measured at threshold is much lower than that predicted by soft pion theorems. This feature has not been understood fully although , recently, *Nath* and *Singh* [9.45] have given an interesting explanation, attributing this to chiral symmetry breaking.

Already, experimental results on the differential cross sections of charged pion photoproduction are becoming available at the resonance region in recent years from the Bates laboratory. Some of the reactions that have been studied recently are

$$^{10}B(\gamma, \pi^+)\,^{10}Be(g.s.) \quad , \qquad ^{14}N(\gamma, \pi^+)\,^{14}C(g.s.) \quad , \qquad \cdot$$

$$^{13}C(\gamma, \pi^-)\,^{13}N(g.s.) \quad , \qquad ^{16}O(\gamma, \pi^+)\,^{16}N(2^-, 0^-, 3^-, 1^-) \quad .$$

Serious discrepancies have been observed between theory and experiment in most cases. Only in the case of ^{16}O, reasonable agreement has been obtained at 320 MeV between the experimental data [9.46] and the theoretical calculation of *Eramzhyan* et al. [9.47] using the momentum space DWIA formalism. The configuration space DWIA calculations of *Decarlo* and *Freed* [9.48] and *Girija* and *Devanathan* [9.49] also yield similar results. The origin of discrepancy in the other cases in the resonance region is not known and a greater effort should be made to investigate the charged pion photoproduction in the resonance region and to settle the issue whether the medium modification of the elementary operator is really important in the resonance region. It is also observed that using a particular set of wave functions, agreement is obtained at lower energy ($E_\gamma = 200$ MeV) in the case of $^{14}N(\gamma, \pi^+)\,^{14}C_{g.s.}$ but underestimates the cross section at higher energy (E_γ = 260 MeV). In the case of $^{10}B(\gamma, \pi^+)\,^{10}Be$, agreement is obtained at E_γ = 360 MeV but the theory predicts a much larger cross section at E_γ = 200 MeV. The situation is really confusing and much greater effort is required to get a clearer picture. It is necessary to measure the differential cross section at various photon energies for several nuclei and compare them with the theoretical inputs. The several terms in the elementary amplitude have different strengths at different energies and the nuclear cross section for a particular transition involves these terms with different weight factors and hence the resulting differential cross section as a function of energy will depend on the delicate combination of these different terms with the corresponding weight factors.

In another two or three years, CW beams with energy ≈ 1 GeV are expected to be operative both at Bates and at NIKHEF. In the year 1993, CEBAF is expected to yield CW beams of energy 4 GeV.

With the prospects of having new facilities of higher energy electron beams with 100 % duty factor, the study of photoproduction and electroproduction of pions and more exotic mesons such as K and η is bound to receive a greater impetus. The possibility of polarization experiments and double coincidence cross section measurements will definitely yield more precise information regarding the detailed structure of the reaction amplitude and the modification of the amplitude in the nuclear medium. If there is a rich variety of experimental possibilities, then an experiment can always be designed to study particular aspects of nuclear structure or particular terms in the reaction amplitude.

The conventional nuclear physics treats protons and neutrons as the constituents of the nucleus and soon pions and deltas have been added to understand the interaction of an external probe with the nucleus. Nuclear physics is rapidly undergoing a renaissance and the nucleus is now being thought of as a composite object of quarks and gluons. There are unambiguous signals for this substructure of the nucleon in high energy lepton scattering experiments. But now myriads of questions arise. How will this new quark-gluon degree of freedom affect the ground state of nuclear matter? Is it possible to explain the scattering and photoproduction of pions at the quark level? Much effort is now being diverted to learn this new physics and the scientific goal of CEBAF, in the words of *Holmgren* [9.42], is "to study the structure of a nuclear many-body system, its quark substructure and the strong and electroweak interactions governing behavior of the fundamental form of matter". We are now at the threshold of excitement of having a panoramic view of this new era of physics that will be ushered in by such high intensity, high quality electron beams.

Appendix

A. The Generalized Helm Model

The generalized Helm model provides a convenient way to parametrize nuclear transition densities and an effecitve tool to determine the multipolarities of individual transitions. Transition densities obtained by analyzing one reaction may readily be transferred to another reaction with similar single particle transition operators. In electron scattering three types of nuclear transition densities contribute to lowest order

$$\varrho(r) = \left\langle J_f M_f \left| \sum_j \frac{1 + \tau_3^j}{2} \delta(r - r_j) \right| J_i M_i \right\rangle \tag{A.1}$$

$$j_c(r) = \left\langle J_f M_f \left| \sum_j \frac{1 + \tau_3^j}{2} \frac{p_j}{m} \delta(r - r_j) \right| J_i M_i \right\rangle \tag{A.2}$$

$$\mu_s(r) = \left\langle J_f M_f \left| \frac{1}{2M} \sum_j \left[\frac{\mu_p + \mu_n}{2} + \frac{\mu_p - \mu_n}{2} \tau_3^j \right] \right. \right.$$
$$\left. \left. \times \sigma^j \delta(r - r_j) \right| J_i M_i \right\rangle \tag{A.3}$$

corresponding to the transition charge, convection current, and spin magnetization densities, respectively. In pion photoproduction the relevant transition matrix elements are

$$\varrho(r) = \left\langle J_f M_f \left| \sum_j \sigma_j \tau_j^\alpha \delta(r - r_j) \right| J_i M_i \right\rangle \tag{A.4}$$

$$\varrho(r) = \left\langle J_f M_f \left| \sum_j \tau_j^\alpha \delta(r - r_j) \right| J_i M_i \right\rangle . \tag{A.5}$$

The electron scattering transition operators contain both isoscalar and isovector components whereas in pion photoproduction only the latter are needed. The Helm model can thus be applied only to those transitions for which it can be established that the isoscalar transition densities vanish or are at least small compared to the isovector densities. For transitions from $T = 0$ to $T = 1$ states, e.g., the isoscalar parts do not contribute at all. Thus in these cases the nuclear transition densities which can be extracted from electron scattering data, can

135

be directly (or possibly after a trivial isospin rotation) used in photoproduction calculations.

In order to analyze reactions involving transitions between discrete states, the appropriate multipoles have to be projected out of the transition densities (A.1–5). The transition charge multipoles are expressed in terms of the reduced matrix elements

$$\varrho_L^{if}(r) = \langle J_f \parallel \tilde{\varrho}_L(r) \parallel J_i \rangle \tag{A.6}$$

where $\tilde{\varrho}_{LM}(r)$ are the multipole components of the transition operator appearing in (A.1). The Fourier–Bessel transform relates (A.6) to the transition charge multipoles

$$\mathcal{M}_L^{if}(q) = (-i)^L \int dr \, r^2 \, j_L(qr) \, \varrho_L^{if}(r) \quad . \tag{A.7}$$

The electric and magnetic transition multipoles are similarly expressed as [A.1]

$$T_L^E(q) = \left[\frac{\sqrt{L+1}}{\hat{L}} \, \mathcal{M}_{LL-1}^j(q) + \frac{\sqrt{L}}{\hat{L}} \, \mathcal{M}_{LL+1}^j(q) \right] - q \, \mathcal{M}_{LL}^\mu(q) \tag{A.8}$$

and

$$T_L^M(q) = \mathcal{M}_{LL}^j(q) + q \left[\frac{\sqrt{L+1}}{\hat{L}} \, \mathcal{M}_{LL-1}^\mu(q) + \frac{\sqrt{L}}{\hat{L}} \, \mathcal{M}_{LL+1}^\mu(q) \right] \tag{A.9}$$

where

$$\mathcal{M}_{LL'}^{j\,if}(q) = (-i)^{L'} \int dr \, r^2 \, j_{L'}(qr) \, j_{LL'}^{if}(r) \tag{A.10}$$

$$\mathcal{M}_{LL'}^{\mu\,if}(q) = (-i)^{L'} \int dr \, r^2 \, j_{L'}(qr) \, \mu_{LL'}^{if}(r) \tag{A.11}$$

are the convection current and magnetization transition density multipoles and where $j_{LL'}^{if}(r)$ and $\mu_{LL'}^{if}(r)$ are the reduced vector multipole components of (A.2) and (A.3) respectively.

Electron scattering cross sections are expressed in terms of the multipoles (A.7–9). Thus by analyzing electron scattering data for a particular transition values for the multipole expressions (A.7–9) can in principle be extracted for the multipolarities contributing to the transition.

The Helm model introduces a description for the multipole components of the transition density matrix elements $(\varrho_L^{if}, j_{LL'}^{if}, \mu_{LL'}^{if})$ which leads to a convenient parametrization of the q-dependence of the transition multipoles $\mathcal{M}_L^{if}(q)$, $\mathcal{M}_{LL'}^{j\,if}(q)$ and $\mathcal{M}_{LL'}^{\mu\,if}(q)$. Extracting the nucleon transition densities from electron scattering data is thus reduced to finding a small set of parameters which provides the best fit to the data.

In the Helm model the transition operators as they are contained in (A.1–5) are assumed to be represented as a convolution of a Gaussian

$$\varrho_g(\boldsymbol{r}) = \frac{\mathrm{e}^{-r^2/g^2}}{(2\pi g)^{3/2}} \tag{A.12}$$

with a δ-function of the form

$$\tilde{\varrho}_R(\boldsymbol{r}) = C\,\frac{\delta(r - R)}{R^2} \tag{A.13}$$

resulting in

$$\tilde{\varrho}(\boldsymbol{r}) = \int \tilde{\varrho}_R(\boldsymbol{r} - \boldsymbol{r}')\,\varrho_g(\boldsymbol{r}')\,d^3r' \quad . \tag{A.14}$$

With this structure for the transition operators, expressions (A.7), (A.10) and (A.11) for the transition density multipoles acquire a factor

$$f_g(q) = \mathrm{e}^{-q^2 g^2/2} \tag{A.15}$$

in front of the integrals, and the density multipoles inside the integrals change to the reduced density multipoles ϱ_{RL}^{if}, $J_{RLL'}^{if}$ and $\mu_{RLL'}^{if}$. For the charge density multipoles this leads to the following form

$$\mathcal{M}_L^{if}(q) = f_g(q)\,(-\mathrm{i})^L \int dr\, r^2\, j_L(qr)\, \varrho_{LR}^{if}(r) \quad . \tag{A.16}$$

For the convection current and magnetization density multipoles similar expressions are obtained.

For the transition charge density and the transition magnetization density the reduced density multipole are chosen to be

$$\varrho_{RL}^{if}(r) = (-\mathrm{i})^{-L}\,\hat{J}_i\,\beta_L\,\frac{\delta(r - R)}{R^2} \tag{A.17}$$

where $\hat{J}_i = (2J_i + 1)^{1/2}$, and

$$\mu_{RLL'}^{if}(r) = \mathrm{i}^{L'}\,\hat{J}_i\,\gamma_{LL'}\,\frac{\delta(r - R)}{2M\overline{R}^2} \quad , \qquad L' = L - 1,\, L,\, L + 1 \quad . \tag{A.18}$$

If (A.17) is inserted into (A.16) the transition charge multipoles reduce to

$$\mathcal{M}_L^{if}(q) = \hat{J}_i\,\beta_L\,f_g(q)\,j_L(qR) \quad . \tag{A.19}$$

Inserting (A.18) into the expression equivalent to (A.16) yields

$$\mathcal{M}_{LL'}^{\mu\,if}(q) = \hat{J}_i\,\frac{\gamma_{LL'}}{2M}\,f_{\overline{g}}(q)\,j_{L'}(q\overline{R}) \quad . \tag{A.20}$$

According to (A.8,9) the $L' = L \pm 1$ components of the transition magnetization density multipoles (A.20) contribute to transverse electric transitions.

The non-rotational part of the transition current density has to satisfy the continuity equation and can thus be related to the transition charge density. Assuming a simple physical model for this part of the current density, and using the continuity equation it is found that [A.1]

$$j^{\text{if}}_{LL-1}(r) = (-\text{i})^{-L} \frac{\hat{J}_i \beta_L}{R^2} \text{i}\omega \frac{\hat{L}}{\sqrt{L}} \left(\frac{r}{R}\right)^{L-1} \begin{cases} 1 & r \leq R \\ 0 & r > R \end{cases} \tag{A.21a}$$

$$j^{\text{if}}_{LL+1}(r) = 0 \tag{A.21b}$$

which leads to the transition multipoles

$$\mathcal{M}^{j\,\text{if}}_{LL-1}(q) = -\omega\,\hat{J}_i\,\beta_L \frac{\hat{L}}{\sqrt{L}}\, f_g(q)\, j_{L-1}(qR) \tag{A.22}$$

$$\mathcal{M}^{j\,\text{if}}_{LL+1}(q) = 0 \quad . \tag{A.23}$$

The L, $L-1$ and L, $L+1$ convection current multipoles contribute to the transverse electric transitions.

The LL current multipoles (corresponding to the rotational part of the transition current) which contribute to transverse magnetic transitions, have thus far not been considered within the framework of the Helm model. In some cases there is evidence that the transition current contribution to magnetic transition is small and that it can thus be ignored. In these cases the Helm model can safely be applied. However, the rotational part of the convection current is certainly not always negligible. To take this component of the current into account a physical model would have to be devised from which then explicit expressions for the reduced current density multipoles $j^{\text{if}}_{LL}(r)$ and the explicit term of $\mathcal{M}^{j\,\text{if}}_{LL}(q)$ could be derived.

According to (A.15) and (A.18) three free parameters are available to fit transition charge multipoles: the transition radius R (i.e. the radius at which the transition is peaked), the width g of the transition region and the strength parameter β_L. Transverse electric transitions are described in terms of 6 parameters: R, g, and β_L according to (A.22) and $\overline{R}$, $\overline{g}$ and γ_{LL} according to (A.20). Fitting form factors of transverse magnetic transitions (assuming $\mathcal{M}^{j\,\text{if}}_{LL} \approx 0$ is a good approximation) requires according to (A.20) four parameters: $\overline{R}$, $\overline{g}$, γ_{LL-1} and γ_{LL+1}.

In pion photoproduction the generalized Helm model is applied in a similar way. Here cross sections are defined in terms of the spinflip and non-spinflip transition density multipoles

$$\varrho^{\text{if}}_{K\gamma}(r) = \langle J_f \parallel \tilde{\varrho}_{K\gamma}(r) \parallel J_i \rangle \tag{A.24a}$$

$$\varrho^{\text{if}}_{K}(r) = \langle J_f \parallel \tilde{\varrho}_{K}(r) \parallel J_i \rangle \tag{A.24b}$$

where $\tilde{\varrho}_{K\gamma}(r)$ and $\tilde{\varrho}_{K}(r)$ are the multipole components of the transition operators appearing in (A.4) and (A.5). In a configuration space calculation photopion cross sections can be expressed as a linear superposition of radial integrals of the type

$$I^{(n)}_{Ll\mathcal{L}\gamma K} = \int dr\, r^2\, \phi^{(n)}_{Ll\mathcal{L}\gamma K}(r)\, \varrho^{(n)}_{K\gamma}(r) \tag{A.25}$$

where L and l specify the photon and pion partial waves, $\mathcal{L}$ the intermediate recoupling between the two, K the transition multipolarity and where $\gamma = K$ for

$n = 0$ (the non-spinflip case) and $\gamma = K \pm 1$ for $n = 1$ (the spinflip case). The photon and pion partial waves as well as the elementary amplitudes are contained in the function ϕ.

When applying the Helm model, the transition densities are represented as in (A.14) as a convolution integral. This results as in the case of the electron scattering multipoles in a Gaussian factor (A.15) appearing in front of the integrals and in the densities under the integrals being replaced by reduced densities (compare (A.16)):

$$I_{LlL\gamma K}^{(n)} = f_g(q) \int dr \, r^2 \, \phi_{LlL\gamma K}^{(n)}(r) \, \varrho_{K\gamma}^{(n)}(r) \quad . \tag{A.26}$$

By inserting again, as for electron scattering, specific expressions for the reduced densities

$$\varrho_{K\gamma R}^{(0)}(r) = (-\mathrm{i})^{-K} \, \hat{\jmath}_i \, \overline{\beta}_K \, \frac{\delta(r - R)}{R^2}$$

$$\varrho_{K\gamma R}^{(i)}(r) = (-\mathrm{i})^{-\gamma} \, \hat{\jmath}_i \, \overline{\gamma}_{K\gamma} \, \frac{\delta(r - R)}{R^2} \tag{A.27}$$

the radial integrals (A.26) reduce to

$$I_{Ll\gamma K}^{(n)} = \begin{cases} (-\mathrm{i})^{-K} \, \hat{\jmath}_i \, \overline{\beta}_K \, \phi_{LlL\gamma K}^{(0)}(R) \, f_g(q) \\ (-\mathrm{i})^{-\gamma} \, \hat{\jmath}_i \, \overline{\gamma}_{K\gamma} \, \phi_{LlL\gamma K}^{(1)}(\overline{R}) \, f_{\overline{g}}(q) \end{cases} \tag{A.28}$$

The photoproduction transition densities $\varrho_K^{(0)}(r)$ and $\varrho_{K\gamma}^{(1)}(r)$ needed in (A.26) are obtained by comparing the transition densities in (A.17) and (A.18) with those of (A.27). Taking into account the differences in the transition operators in (A.11) vs (A.5) and in (A.3) vs (A.4) one obtains

$$\varrho_K^{(0)}(r) = \sqrt{2} \, \varrho_K^{\mathrm{if}}(r)$$

$$\varrho_{K\gamma}^{(1)}(r) = \frac{\sqrt{2}}{\mu_p - \mu_n} \, \varrho_{K\gamma}^{\mathrm{if}}(r) \tag{A.29}$$

which expresses the photoproduction transition densities in terms of the electron scattering transition densities. The factor $\sqrt{2}$ is appropriate for charged pion production. It accounts for the difference in the nucleon isospin transition operator in the two reactions. From (A.29) it follows that the strength parameters β and $\gamma_{K\gamma}$ derived from electron scattering are related to the strength parameters for photoproduction as

$$\overline{\beta}_K = \sqrt{2} \, \beta_K$$

$$\overline{\gamma}_{K\gamma} = \frac{\sqrt{2}}{\mu_p - \mu_n} \, \gamma_{K\gamma} \quad .$$

Since in (A.25) the explicit evaluation of the integrals has been eliminated, the computational effort is considerably reduced. However, the results may not always be reliable. A better approach is to use the inverses of the transformations (A.7) and (A.16). Inserting into these multipoles (A.19) and (A.20) one obtains the electron scattering transition densities in configuration space

$$\varrho_L^{if}(r) = \frac{2}{\pi} i^L \hat{J}_i \beta_L \int dq \, q^2 \, j_L(qr) \, j_L(qR) e^{-g^2 q^2/2}$$

$$\tag{A.30}$$

$$\varrho_{LL'}^{if}(r) = \frac{2}{\pi} i^{L'} \hat{J}_i \frac{\gamma_{LL'}}{2M} \int dq \, q^2 \, j_{L'}(qr) \, j_{L'}(q\overline{R}) e^{-\overline{g}^2 q^2/2} \quad .$$

The transition densities to be used in the integrals (A.26) are then obtained by simply inserting (A.30) into (A.29).

References

Chapter 1

1.1 A. Nagl: "Charged pion photoproduction on light nuclei", Ph.D. Thesis, Catholic University of America (1983)

1.2 V. Girija: "Photoproduction of Pions from Nuclei in Distored Wave Impulse Approximation", Ph.D. Thesis, University of Madras (1982)

1.3 M.K. Singham: "Charged Pion Photoproduction from Light Nuclei", Ph.D. Thesis, University of Pittsburgh (1980)

1.4 P. Bosted: "Pion Photoproduction in the (3,3) Resonance Region", Ph.D.Thesis, M.I.T. (1980)

1.5 P. Stoler (ed.): *Photopion Nuclear Physics* (Plenum, New York/London 1979)

1.6 V. Devanathan and P.R. Subramanian (eds.): *"Medium Energy Physics and Nuclear Structure*, J. of Madras University, Sec. B, Vol. 45 (1982)

1.7 V. Devanathan and P.R. Subramanian (eds.): *Perspectives in Nuclear Physics*, J. of Madras University, Sec. B, Vol. 50 (1987)

1.8 A. Nagl: "Nuclear Pion Photoproduction", in *Perspectives in Nuclear Physics*, ed. by V. Devanathan and P.R. Subramanian (University of Madras 1987) pp. 410–450

1.9 A.M. Bernstein: "An Introduction to (γ, π) Reactions" in: Proc. International School of Intermediate Energy Nuclear Physics, Verona, Italy, 1981, ed. by R. Bergère, C. Costa and C. Schaerf (World Publishing Co., Singapore 1982)

1.10 A.M. Bernstein: "Electromagnetic Meson Production". Invited Talk at the Seminar on Electromagnetic Interactions of Nuclei at Low and Medium Energies, Moscow, USSR, Dec. 1988

1.11 J.M. Laget: Physics Reports **69**, 1 (1981)

1.12 M.K. Singham, F. Tabakin: Ann. Phys. (N.Y.) **135**, 71 (1981)

1.13 A.A. Chumbalov, R.A. Eramzhyan, S.S. Kamalov: Report E4–88–917 (1988), E4–89–11 (1989), Institute for Nuclear Research of the USSR Academy of Sciences, Moscow, USSR

1.14 K. Shoda: "Photoproduction of Mesons on Nuclei" in: Proc. of the 6th Seminar on Electromagnetic Interactions of Nuclei at Low and Medium Energies, Moscow, Dec. 10–12, 1984

1.15 E.W. Laing, R.G. Moorhouse: Proc. Phys. Soc. **70**, 629 (1957)

1.16 V. Devanathan, G. Ramachandran: Nucl. Phys. **38**, 654 (1962); **42**, 25 (1963); **66**, 595 (1965)

1.17 V. Devanathan, M. Rho, K.S. Rao, S.C.K. Nair: Nucl. Phys. **B2**, 329 (1967)

1.18 H. Überall: *Electron Scattering from Complex Nuclei* (Academic Press, New York, 1971)

1.19 H. Überall: Suppl. Nuovo Cimento **4**, 781 (1966)

1.20 J.D. Walecka: Proc. of the Williamsburg Conf. on Intermediate Energy Physics, College of William and Mary, Williamsburg, Virginia, 1966

1.21 K. Shoda, H. Ohashi, K. Nakahara: Phys. Rev. Lett. **39**, 1131 (1977)

Chapter 2

2.1 E.H. Bellamy: Prog. Nucl. Phys. **8**, 237 (1960)

2.2 H.A. Bethe, F. de Hoffmann: *Mesons and Fields*, Vol.II (Row, Peterson and Co., Evenston IL 1955)

2.3 E.M. McMillan et al.: Science **109**, 438 (1949)

2.4   J. Steinberger, A.S. Bishop: Phys. Rev. **78**, 493 (1950)

2.5   R.M. Littauer, D. Walker: Phys. Rev. **86**, 838 (1952)

2.6   R.R. Wilson: Phys. Rev. **87**, 125 (1952)

2.7   M. Lax, H. Feshbach: Phys. Rev. **81**, 189 (1951)

2.8   S.T. Butler: Phys. Rev. **87**, 1117 (1952)

2.9   J.R. Waters: Phys. Rev. **113**, 1133 (1959)

2.10 G.F. Chew: Phys. Rev. **80**, 196 (1950);
      G.F. Chew, G.C. Wick: Phys. Rev. **85**, 636 (1952)

2.11 J. Steinberger, A.S. Bishop: Phys. Rev. **86**, 171 (1952)

2.12 J. Steinberger, W.K.H. Panofsky, J.S. Stellar: Phys. Rev. **78**, 802 (1950)

2.13 W.K.H. Panofsky, J. Steinberger, J.S. Stellar: Phys. Rev. **86**, 180 (1952)

2.14 K.M. Watson: Phys. Rev. **95**, 228 (1954)

2.15 G.F. Chew, M.L. Goldberger, F.E. Low, Y. Nambu: Phys. Rev. **106**, 1345 (1957)

2.16 V. Devanathan, G. Ramachandran: Nucl. Phys. **23**, 312 (1961)

2.17 A. Ramakrishnan, V. Devanathan, G. Ramachandran: Nucl. Phys. **24**, 163 (1961)

2.18 V. Devanathan, G. Ramachandran: Nucl. Phys. **38**, 654 (1962)

2.19· V. Devanathan, G. Ramachandran: Nucl. Phys. **42**, 256 (1963)

2.20 I.S. Hughes, P.V. March: Proc. Phys. Soc. (London) **72**, 259 (1958)

2.21 P. Dyal, J.P. Hummel: Phys. Rev. **127**, 2217 (1962)

2.22 P.V. March, J.G. Walker: Proc. Phys. Soc. (London) **77**, 293 (1960)

2.23 R.A. Meyer, W.B. Walters, J.P. Hummel: Phys. Rev. **B138**, 1421 (1965)

2.24 G. Nydahl, B. Forkman: Nucl. Phys. **B7**, 97 (1968)

2.25 G. Ramachandran, V. Devanathan: Nucl. Phys. **66**, 595 (1965)

2.26 V. Devanathan, M. Rho, K.S. Rao, S.C.K. Nair: Nucl. Phys. **B2**, 329 (1967)

2.27 V. Devanathan, K.S. Rao, R. Sridhar: Phys. Lett. **25B**, 456 (1967)

2.28 K.S. Rao, V. Devanathan, G.N.S. Prasad: Nucl. Phys. **A159**, 97 (1970);
      V. Devanathan: "Effects of short-range correlations in some reaction processes", in: Symp. on Theor. Phys., Vol. 10, ed. by A. Ramakrishnan (Plenum Press, N.Y. 1970) pp. 115;
      V. Devanathan: "Nuclear distortion effects on the emitted pion in photopion reactions", in: Symp. on Theor. Phys. Vol. 10, ed. by A. Ramakrishnan (Plenum Press, N.Y. 1970) pp. 123;
      V. Devanathan, G.N.S. Prasad, K.S. Rao: Phys. Rev. **C8**, 18 (1973)

2.29 L.M. Saunders: Nucl. Phys. **B7**, 293 (1968)

2.30 F.A. Berends, A. Donnachie, D.L. Weaver: Nucl. Phys. **B4**, 1 (1967)

2.31 P. Pascual, J.L. Sanchez–Gomez: Nucl. Phys. **B16**, 191 (1970);
      D. Griffith, C.W. Kim: Nucl. Phys. **B6**, 49 (1968)

2.32 F.J. Kelly, L.J. McDonald, H. Überall: Nucl. Phys. **A139**, 329 (1969)

2.33 H. Überall, B.A. Lamers, C.W. Lucas, A. Nagl: Phys. Lett. **44B**, 324 (1973)

2.34 J.H. Koch, T.W. Donnelly: Nucl. Phys. **B64**, 478 (1973)

2.35 J.B. Seaborn, V. Devanathan, H. Überall: Nucl. Phys. **A219**, 461 (1974)

2.36 F. Cannata et al.: Can. J. Phys. **52**, 1405 (1974)

2.37 K. Stricker, H. McManus, J.A. Carr: Phys. Rev. **C19**, 929 (1979);
      K. Stricker, J.A. Carr, H. McManus: Phys. Rev. **C22**, 2043 (1980);
      J.A. Carr, H. McManus, K. Stricker–Bauer: Phys. Rev. **C25**, 952 (1982)

2.38 A. Nagl, H. Überall: Phys. Lett. **B96**, 254 (1980);
      A. Nagl, H. Überall: Phys. Lett. **B63**, 291 (1976)

2.39 M.K. Singham, F. Tabakin: Phys. Rev. **C21**, 1039 (1980)
      M.K. Singham, F. Tabakin: Ann. Phys. (N.Y.) **135**, 71 (1981)

2.40 V. DeCarlo et al.: Phys. Rev. **C21**, 1460 (1980);
      N. Freed, W. Rhodes: Lett. Nuovo Cimento **18**, 88 (1977);
      N. Freed, P. Ostrander: Phys. Rev. **C11**, 805 (1975);
      N. Freed, P. Ostrander: Phys. Lett. **B61**, 449 (1976);
      K.I. Blomqvist, P. Jaecek, G.G. Jonsson, H. Dinter, K. Tesch, N. Freed, P. Ostrander: Phys. Rev. **C15**,988 (1977)

2.41 V. Girija, V. Devanathan: Phys. Rev. **26**, 2152 (1982)

2.42 L. Tiator, L.E. Wright: Phys. Rev. **C30**, 989 (1984)

2.43 A. Figureau, N.C. Mukhopadhyay: Nucl. Phys. **A338**, 514 (1980)

2.44 R. Wittman, R. Davidson, N.C. Mukhopadhyay: Phys. Lett. **142B**, 336 (1984);
     R. Wittman, N.C. Mukhopadhyay: Phys. Rev. Lett. **57**, 1113 (1986)

2.45 I. Blomqvist, J.M. Laget: Nucl. Phys. **A280**, 405 (1977)

2.46 G. Toker, F. Tabakin: Phys. Rev. **C28**, 1725 (1983)

2.47 S.A. Dytman, F. Tabakin: Phys. Rev. **C33**, 1699 (1986)

2.48 G. Audit et al.: Phys. Rev. **C15**, 1415 (1977)

2.49 E.C. Booth et al.: Phys. Lett. **66B**, 236 (1977)

2.50 G. Audit et al.: Phys. Rev. **C16**, 1517 (1977)

2.51 F.L. Milder et al.: Phys. Rev. **C19**, 1416 (1979)

2.52 E.C. Booth et al.: Phys. Lett. **C20**, 1603 (1979)

2.53 A.M. Bernstein et al.: Phys. Rev. Lett. **37**, 819 (1976)

2.54 P.E. Bosted, K.I. Blomqvist, A.M. Bernstein: Phys. Rev. Lett. **43**, 1473 (1979)

2.55 N. Paras et al.: Phys. Rev. Lett. **42**, 1455 (1979)

2.56 V.D. Epaneshnikov et al.: Sov. J. Nucl. Phys. **19**, 242 (1974)

2.57 K. Min et al.: Phys. Rev. **C14**, 807 (1976)

2.58 K. Min et al.: Phys. Rev. Lett. **44**, 1384 (1980)

2.59 P.E. Bosted et al.: Phys. Rev. Lett. **45**, 1544 (1980)

Chapter 3

3.1 H. Genzel et al.: Landolt–Börnstein, Numerical data and functional relationships in science
    and technology, Vol. 8 (Springer, Berlin, 1973);
    T. Fujii et al.: Nucl. Phys. **B120**, 395 (1977);
    G. Von Holtey: Springer Tracts in Modern Physics **59**, 3 (1971)

3.2 J.M. Laget: Phys. Reports **69**, 1 (1981)

3.3 K.I. Blomqvist, J.M. Laget: Nucl. Phys. **A280**, 405, (1977)

3.4 A.B.B.H.H.M Collaboration: Phys. Rev. **175**, 405 (1977)

3.5 P. Benz et al.: Nucl. Phys. **B65**, 158 (1977);
    T. Fujii et al.: Nucl. Phys. **B120**, 395 (1977); Phys. Rev. Lett. **28**, 672 (1972) and **29**, 244
    (1972)

3.6 K.M. Watson: Phys. Rev. **95**, 228 (1954)

3.7 H. Tanabe, K. Ohta: Phys. Rev. **C31**, 1876 (1985)

3.8 G.F. Chew, M.L. Goldberger, F.E. Low, Y. Nambu: Phys. Rev. **106**, 1345 (1957)

3.9 G. Höhler, A. Müllensiefen: Z. Physik **157**, 30 (1957);
    G. Höhler, K. Dietz, A. Müllensiefen: Z. Physik **159**, 77 (1960)
    G. Höhler, W. Schmidt, Ann. Phys. **28**, 34 (1967)

3.10 F.A. Berends, A. Donnachie, D.L. Weaver: Nucl. Phys. **B4**, 1 (1967) and **B4**, 57 (1967)

3.11 J.D. Bjorken, S.D. Drell: Relativistic Quantum Mechanics (McGraw Hill, New York, 1964)

3.12 M.G. Olsson, E.T. Osypowski: Nucl. Phys. **B87**, 399 (1975)

3.13 N.M. Kroll, M.A. Ruderman: Phys. Rev. **93**, 233 (1954)

3.14 J.M. Laget: Phys. Rev. Lett. **41**, 89 (1978)

3.15 S. Weinberg: Phys. Rev. **166**, 1568 (1968)

3.16 S. Weinberg: Phys. Rev. Lett. **18**, 188 (1967)

3.17 B.W. Lee, H.T. Nieh: Phys. Rev. **166**, 1507 (1968)

3.18 R.D. Peccei: Phys. Rev. **181**, 1902 (1969)

3.19 M.G. Olsson: Nucl. Phys. **B78**, 55 (1974)

3.20 P. Bosted, J.M. Laget: Nucl. Phys. **A296**, 413 (1978)

3.21 M.K. Singham, F. Tabakin: Ann. Phys. (N.Y.) **135**, 71 (1981);
     L. Tiator, L.E. Wright: Phys. Rev. **C30** , 989 (1984);
     S.A. Dytman, F. Tabakin: Phys. Rev. **C33**, 1699 (1986)

3.22 R. Cenni, G. Dillon, P. Christillin: Nuovo Cim. **97**, 9 (1987);
J.L. Sabutis: Phys. Rev. **C27**, 778 (1983);
R. Wittman, R. Davidson, N.C. Mukhopadhyay: Phys. Lett. **B142**, 336 (1984)
3.23 J.M. Laget: Nucl. Phys. **A481**, 765 (1988)
3.24 H. Tanabe, K. Ohta: Phys. Rev. **C31**, 1876 (1985)
3.25 M. Araki, I.R. Afnan: Phys. Rev. **C36**, 250 (1987)
3.26 T. Wittman, N.C. Mukhopadhyay: Phys. Rev. Lett. **57**, 1113 (1986)
3.27 C. Bennhold, L. Tiator, L.E. Wright: Univ. of Mainz preprint
3.28 L. Tiator et al.: Nucl. Phys. **A485**, 565 (1988)
3.29 T. Suzuki, T. Takaki, J.H. Koch: Nucl. Phys. **A469**, 607 (1986)
3.30 K.H. Althoff: "Polarization experiments", in: *From Collective States to Quarks in Nuclei* (H. Arenhövel and A.M. Saruis, eds.) Lecture Notes in Physics, Vol. 137, Springer-Verlag, Berlin 1981
3.31 I.M. Barbour, R.L. Crawford, N.H. Parsons: Nucl. Phys. **B141**, 253 (1978)
3.32 R. Kajikawa: "Pion Photoproduction and Compton Scattering in the Resonance Region", in: Proceedings of the 4th Int. Conf. on Baryon Resonances (Univ. of Toronto, Toronto 1980), p. 352
3.33 W.J. Metcalf, R.L. Walker: Nucl. Phys. **B76**, 253 (1974)
3.34 P. Noelle: Prog. Theor. Phys. **60**, 778 (1978)
3.35 S. Arai: in: Proceedings of the 4th Int. Conf. on Baryon Resonances (Univ. of Toronto, Toronto 1980), p. 93
3.36 D. Schwela, R. Weizel: Z. Physik **221**, 71 (1969)
3.37 P. Noelle, W. Pfeil, D. Schwela: Nucl. Phys. **B26**, 461 (1971)
3.38 W. Pfeil, D. Schwela: Nucl. Phys. **B45**, 379 (1972)
3.39 R.G. Moorhouse, H. Oberlack, A.H. Rosenfeld: Phys. Rev. **D9**, 1 (1974)
3.40 R.L. Walker: Phys. Rev. **182**, 1729 (1969)
3.41 P. Feller et al.: Nucl. Phys. **B104**, 219 (1976)
3.42 R.L. Crawford, W.T. Morton: Nucl. Phys. **B211**, 1 (1983)
3.43 I.S. Barker, A. Donnachie, J.K. Storrow: Nucl. Phys. **B79**, 431 (1974)
3.44 N. Isgur, G. Karl: Phys. Rev. **D19**, 2653 (1979)
3.45 M. Araki, I.R. Afnan: Phys. Rev. **C36**, 250 (1987)
3.46 S. Scherer, D. Drechsel, L. Tiator: Phys. Lett. **B193**, 1 (1987)
3.47 G. von Holtey et al.: Z. Physik **259**, 51 (1973)
3.48 K.H. Althoff et al.: Nucl. Phys. **B116**, 253 (1976)
3.49 I.S. Barber, A. Donnachie, J.K. Storrow: Nucl. Phys. **B79**, 431 (1974)
3.50 I.S. Barker, A. Donnachie, J.K. Storrow: Nucl. Phys. **B95**, 347 (1975)
3.51 P. Hampe: Report BONN−IR−80−1, Institute of Physics, Bonn University, 1980
3.52 H. Überall: Phys. Rev. **103**, 1055 (1956)
3.53 H. Überall: Z. Naturforsch. **17a**, 332 (1962)
3.54 K.H. Althoff et al.: Z. Physik **C18**, 199 (1983)
3.55 K.H. Althoff et al.: Z. Physik **C1**, 327 (1979)
3.56 K.H. Althoff et al.: Nucl. Phys. **B53**, 9 (1973)
3.57 K.H. Althoff et al.: Phys. Letters **59B**, 93 (1975)
3.58 K.H. Althoff et al.: Phys. Letters **63B**, 107 (1976)
3.59 M. Fukushima et al.: Nucl. Phys. **B130**, 486 (1977)
3.60 K.H. Althoff et al.: Nucl. Phys. **B131**, 1 (1977)
3.61 P. Noelle: Report BONN−IR−75−20, Institute of Physics, Bonn University, 1976
3.62 K.H. Althoff et al.: Nucl. Phys. **B96**, 497 (1975)
3.63 K.H. Althoff et al.: Nucl. Phys. **B116**, 253 (1976)
3.64 H. Becks et al.: Nucl. Phys. **B60**, 267 (1973)
3.65 E. Hilger et al.: Z. Physik **268**, 19 (1974)
3.66 K. Jaeckel et al.: Z. Physik **268**, 27 (1974)
3.67 H. Genzel: Z. Physik **268**, 37 (1974)

3.68 H. Genzel et al.: Z. Physik **268**, 43 (1974)
3.69 P. Feller et al.: Phys. Letters **49B**, 197 (1974)
3.70 W. Brefeld et al.: Nucl. Phys. **B100**, 93 (1975)
3.71 P. Blüm et al.: Z. Physik **A277**, 311 (1976)
3.72 P. Büm et al.: Z. Physik **A278**, 275 (1976)
3.73 H. Herr et al.: Nucl. Phys. **B125**, 157 (1977)
3.74 D. Husmann et al.: Nucl. Phys. **B126**, 436 (1977)
3.75 H. Breuker et al.: Nucl. Phys. **B146**, 285 (1978)
3.76 H. Breuker et al.: Z. Physik **C13**, 113 (1982)
3.77 H. Breuker et al.: Z. Physik **C17**, 121 (1983)
3.78 K. Bätzner et al.: Nucl. Phys. **B76**, 1 (1974)
3.79 H. Blume et al.: Z. Physik **C16**, 283 (1983)
3.80 E. Amaldi, S. Fubini, G. Furlan: *Pion-Elelctroproduction*, Springer Tracts in Modern Physics Vol. 83 (Springer-Verlag Berlin, Heidelberg 1979)
3.81 W. Schmidt, G. Schwiderski: Forts. Physik **15**, 393 (1967)
3.82 Particle Data Group: Phys. Lett. **204B**, 377 (1988)

Chapter 4

4.1 M.E. Rose: *Elementary Theory of Angular Momentum* (John Wiley, New York, 1957)
4.2 K.S. Rao, V. Devanathan: Phys. Lett. **32B**, 578 (1970)
4.3 K.S. Rao, V. Devanathan: Can. J. Phys. **53**, 1299 (1975)
4.4 A. Nagl, F. Cannata, H. Überall et al.: Phys. Rev. **C12**, 1586 (1975)
4.5 H. Überall et al.: Phys. Lett. **44B**, 324 (1973)
4.6 K.S. Rao: Phys. Rev. **C7**, 1785 (1973)
4.7 K.S. Rao: J. Phys. **A4**, 924 (1971)
4.8 A.K. Rej, T. Engeland: Phys. Lett. **45B**, 77 (1973)
4.9 V. Gillet, N. Vinh Mau: Nucl. Phys. **54**, 321 (1964)
4.10 M. Rho: Phys. Rev. Lett. **18**, 671 (1967); Phys. Rev. **161**, 955 (1967)
4.11 J.L. Friar, B.F. Gibson: Phys. Rev. **C15**, 1779 (1977)
4.12 P.K. Teng et al.: Phys. Lett. **B177**, 25 (1986)
4.13 M. Hirata, F. Lenz, K. Yazaki: Ann. Phys. **108**, 116 (1977)
4.14 G. Chanfray, M. Ericson: Phys. Lett. **141B**, 163 (1984);
       G. Chanfray, J. Delorme: Phys. Lett. **129B**, 167 (1983)
4.15 N.C. Mukhopadhyay, H. Toki, W. Weise: Phys. Lett. **84B**, 35 (1979)
4.16 J. Cohen, J.M. Eisenberg: Phys. Rev. **C28**, 1309 (1983);
       J. Cohen: Phys. Rev. **C30**, 1238; **C30**, 1573 (1985); **C29**, 914 (1985)
4.17 L.S. Kisslinger, W.L. Wang: Phys. Lett. **30**, 1071 (1973); Ann. Phys. (N.Y.) **99**, 374 (1976)
4.18 M. Hirata, J.H. Koch, F. Lenz, E.J. Moniz: Ann. Phys. (N.Y.) **120**, 205 (1979)
4.19 J.H. Koch, E.J. Moniz: Phys. Rev. **C27**, 751 (1983)
4.20 T. Takaki, T. Suzuki, J.H. Koch: Nucl. Phys. **A443**, 570 (1985)
4.21 T. Suzuki, T. Takaki, J.H.Koch: Nucl. Phys. **A460**, 607 (1986)
4.22 B.H. Cottman et al.: Phys. Rev. Lett. **55**, 684 (1985)

Chapter 5

5.1 R.A. Eisenstein, F. Tabakin: Comp. Phys. Commun. **17**, 23 (1976)
5.2 R.H. Landau, S.C. Phatak, F, Tabakin: Ann. Phys. (N.Y.) **78**, 299 (1973); Phys. Rev. **C7**, 1803 (1973)
5.3 R.S. Bhalerao, L.C. Liu, C.M. Shakin: Phys. Rev. **C21**, 1903 (1980)
5.4 R.A. Eisenstein, G.A. Miller: Comp. Phys. Commun. **8**, 130 (1974)
5.5 L.S. Kisslinger: Phys. Rev. **98**, 761 (1955)

5.6   G. Fäldt: Nucl. Phys. **A206**, 176 (1973)
5.7   M. Krell, S. Barmo: Nucl. Phys. **B20**, 461 (1970)
5.8   M.A. Melkanoff et al.: "A Fortran Program for elastic scattering analysis with the nuclear optical model" (University of California Press 1962)
5.9   D.S. Koltun: Advances in Nuclear Physics 3, 71 (1969)
5.10   E. Boschitz: Proc. of the Zurich Conference on Intermediate Energy Nuclear Physics (1977)
5.11   B.M. Preedom et al.: Phys. Rev. **C23**, 1134 (1981);
       B.M. Preedom: Proc. of the Zurich Conference on Intermediate Energy Nuclear Physics (1977)
5.12   F. Binon et al.: Nucl. Phys. **B17**, 168 (1970)
5.13   F. Binon et al.: Phys. Rev. Lett. **35**, 145 (1975);
       R.W. Bercaw et al.: Phys. Rev. Lett. **29**, 1031 (1972)
5.14   M. Blecher et al.: Phys. Rev. **C10**, 2247 (1974)
5.15   J.P. Albanese et al.: Nucl. Phys. **A350**, 301 (1980); Phys. Lett. **B73**, 119 (1978)
5.16   Q. Ingram et al.: Phys. Lett. **76B**, 173 (1978)
5.17   J.F. Amann et al.: Phys. Rev. Lett. **35**, 426 (1975)
5.18   S.A. Dytman et al.: Phys. Rev. Lett. **38**, 1059 (1977); **39**, 53(E) (1977)
5.19   S.A. Dytman et al.: Phys. Rev. **C19**, 971 (1979)
5.20   R.R. Johnson et al.: Nucl. Phys. **A296**, 444 (1978)
5.21   R.R. Johnson et al.: Phys. Lett. **78B**, 560 (1978)
       K.L. Erdman et al.: Proc. of the Zurich Conference on Intermediate Energy Nuclear Physics (1977)
5.22   F. Milder et al.: Proc. of the Zurich Conference on Intermediate Energy Nuclear Physics (1977)
5.23   B.D. Keister: Nucl. Phys. **A350**, 365 (1980)
5.24   R. Subramanian, P. Ratna Prasad, V. Devanathan: Pramana (to be published)
5.25   R. Wittman, N.C. Mukhopadhyay: Phys. Rev. Lett. **57**, 1113 (1986)
5.26   J.H. Koch, E.J. Moniz: Phys. Rev. **C27**, 751 (1983)
5.27   A.N. Saharia, R.M. Woloshyn: Phys. Rev. **C23**, 351 (1981)

Chapter 6

6.1   M.K. Singham: Phys. Rev. **C33**, 2194 (1986)
6.2   R.L. Huffman, J. Dubach, R.S. Hicks, M.A. Plum: Phys. Rev. **C35**, 35 (1987)
6.3   S. Cohen, D. Kurath: Nucl. Phys. **73**, 1 (1965)
6.4   R.H. Helm: Phys. Rev. **104**, 1466 (1966)
6.5   M. Rosen, R. Raphael, H. Überall: Phys. Rev. **163**, 927 (1967)
6.6   H. Überall: Springer Tracts in Modern Physics **49**, 1 (1969)
6.7   H. Überall et al.: Phys. Rev. **C6**, 1911 (1972)
6.8   R.D. Graves et al.: Can. J. Phys. **58**, 48 (1980)
6.9   K. Rohrich et al.: Phys. Lett. **153B**, 203 (1985)
6.10   N. Ensslin et al.: Phys. Lett. **153B**, 203 (1985)
6.11   R.L. Huffman et al.: Phys. Lett. **139B**, 203 (1984)

Chapter 7

7.1   I. Blomqvist, G. Nydhal, B. Forkman: Nucl. Phys. **A162**, 193 (1971)
7.2   K. Shoda, H. Ohashi, K. Nakahara: Nucl. Phys. **A350**, 377 (1980)
7.3   K. Shoda, M. Yamazaki, M. Torikoshi, H. Tsubota, K. Min, E.J. Winhold, A.M. Bernstein: Phys. Rev. **C27**, 443 (1983)
7.4   K. Shoda, O. Sasaki, T. Kohimura: Phys. Lett. **101B**, 124 (1981)
7.5   K. Shoda, M. Yamazaki, M. Torikoshi, O. Sasaki, H. Tsubota, B.N. Sung: Nucl. Phys. **A403**, 469 (1983)

7.6   K. Shoda, M. Yamazaki, M. Torikoshi, O. Sasaki, H. Tsubota, B.N. Sung: Jr. Phys. Soc. Japan **52**, 3355 (1983)

7.7   K. Min, E.J. Winhold, K. Shoda, H. Tsubota, H. Ohashi, M. Yamazaki: Phys. Rev. Lett. **44**, 1384 (1980)

7.8   K. Shoda, T. Kobayashi, O. Sasaki, S. Toyama: Nucl. Phys. **A486**, 526 (1988)

7.9   B.N. Sung, K. Shoda, A. Gagaya, S. Toyama, M. Torikoshi, O. Sasaki, T. Kobayashi, H. Tsubota: Nucl. Phys. **A473**, 705 (1987)

7.10   K. Shoda, M. Torikoshi, M. Yamazaki, O. Sasaki, H. Tsubota: Nucl. Phys. **A439**, 669 (1985)

7.11   K. Min, E.J. Winhold, K. Shoda, M. Torikoshi, M. Yamazaki, O. Sasaki, H. Tsubota, B.N. Sung: Phys. Rev. **C28**, 464 (1983)

7.12   K. Shoda, S. Toyama, M. Torikoshi, O. Sasaki, T. Kobayashi: Nucl. Phys. **A48**, 512 (1988)

7.13   B.H. Cottman et al.: Phys. Rev. Lett. **55**, 68 (1985)

7.14   A. Liesenfeld et al.: Nucl. Phys. **A485**, 580 (1988)

7.15   Ch. Schmitt et al.: Nucl. Phys. **A395**, 435 (1983)

7.16   J. LeRose et al.: Phys. Rev. **C26**, 2554 (1982)

7.17   K. Shoda et al.: Phys. Lett. **169B**, 17 (1986)

7.18   P. Stoler et al.: Phys. Lett. **143B**, 69 (1984)

7.19   K. Rohrich et al.: Phys. Lett. **153B**, 203 (1985)

7.20   M. Yamazaki, K. Shoda, M. Torikoshi, O. Sasaki, H. Tsubota: Phys. Rev. **C34**, 1123 (1986)

7.21   K. Shoda et al.: Phys. Rev. **C33**, 2179 (1986)

7.22   M. Yamazaki et al.: Phys. Rev. **C35**, 355 (1987)

7.23   V. Girija, M. Manimekalai, V. Devanathan: Physica Scripta **27**, 398 (1983)

7.24   M. Hirooka, T. Konishi, R. Morita, H. Narumi, M. Soga, M. Morita: Prog. Theor. Phys. **40**, 808 (1968)

7.25   N. Paras, A. Bernstein, K.I. Blomqvist, G. Franklin, M. Pauli, B. Schoch, J. LeRose, K. Min, D. Rowley, P. Stoler, E.J. Winhold, P.F. Yergin: Phys. Rev. Lett. **42**, 1455 (1979)

7.26   R.A. Eramzhyan, M. Gmitro, S.S. Kamalov, R. Mach: J. Phys. **G9**, 605 (1983); R.A. Eramzhyan, M. Gmitro, S.S. Kamalov: Phys. Lett. **128B**, 371 (1983)

7.27   N. Ensslin et al.: Phys. Rev. **C9**, 1705 (1974)

7.28   R. Wittman, N.C. Mukhopadhyay: Phys. Rev. Lett. **57**, 1113 (1986)

7.29   M.K. Singham, F. Tabakin: Ann. Phys. (N.Y.) **135**, 71 (1981)

7.30   S. Malecki: Nucl. Phys. **A403**, 607 (1983)

7.31   M.K. Singham: Phys. Rev. Lett. **54**, 1642 (1985)

7.32   P.S. Hauge, S. Maripuu: Phys. Rev. **C8**, 1609 (1973)

7.33   R.S. Hicks et al.: Phys. Rev. **C26**, 339 (1982)

7.34   J. LeRose et al.: Phys. Rev. **C25**, 1702 (1982)

7.35   L. Tiator: Phys. Lett. **125B**, 367 (1983)

7.36   P. Stoler et al.: Phys. Rev. Lett. **143B**, 69 (1984)

Chapter 8

8.1   L.S. Kisslinger, W.L. Wang: Phys. Rev. Lett. **30**, 1071 (1973)

8.2   L.S. Kisslinger, W.L. Wang: Ann. Phys. (N.Y.) **99**, 374 (1976)

8.3   R.M. Woloshyn: Phys. Rev. **C18**, 1056 (1978)

8.4   A.N. Saharia, R.M. Woloshyn: Phys. Rev. **C23**, 351 (1981)

8.5   J.H. Koch, E.J. Moniz: Phys. Rev. **C20**, 235 (1979)

8.6   J.H. Koch, E.J. Moniz: Phys. Rev. **C27**, 751 (1983)

8.7   T. Takaki, T. Suzuki, J.H. Koch: Nucl. Phys. **A443**, 570 (1985)

8.8   V. Girija, V. Devanathan, A. Nagl, H. Überall: Phys. Rev. **C27**, 1169 (1983)

8.9   S.S. Kamalov, T.D. Kaipov: Phys. Lett. **B162**, 260 (1985)

8.10   S. Boffi, R. Mirando: Nucl. Phys. **A448**, 637 (1986)

8.11   J.E. Leiss, S. Penner, R.A. Schrack: Phys. Rev. **127**, 1772 (1962)

8.12   G. Davidson: Ph. D. thesis, Massachusetts Institute of Technology (1959)

8.13   J.W. Staples: Ph. D. thesis, University of Illinois (1969)
8.14   B. Bellinghausen, A. Christ, H.J. Gassen, G. Goerigk, R. Müller, G. Nöldeke, T. Reichelt, H. Stanek, P. Stipp: Z. Phys. **A309**, 65 (1982)
8.15   J. Comuzzi, R. Becker, E.C. Booth, W.J. Burger, G.W. Dodson, J.P. Miller, R.P. Redwine, B.L. Roberts, D.R. Tieger: Bull. Am. Phys. Soc. **27**, 4 (1982); Phys. Rev. Lett. **53**, 755 (1984)
8.16   J. Arends et al.: in: Abstracts of contributed papers to the IX International Conference on High Energy Physics and Nuclear Structure, Versailles (1981);
       J. Arends, N. Floss, A. Hegerath, B. Mecking, G. Nöldeke, R. Stenz: Z. Phys. **A311**, 367 (1983)
8.17   G. Tamas: Nucl. Phys. **A446**, 327 (1985)
8.18   J. Arends, P. Detemple, N. Floss, A. Hegerath, S. Huthmacher, B. Mecking, G. Nöldeke, R. Stenz, V. Werler: Nucl. Phys. **A454**, 579 (1986)
8.19   J.H. Koch, E.J. Moniz, N. Ohtsuka: Ann. Phys. (N.Y.) **154**, 99 (1984)
8.20   J.H. Koch: Proceedings of the Los Alamos Workshop on "Photon and neutral meson physics at intermediate energies" (1987)
8.21   S.S. Hanna: Proceedings of the Los Alamos Workshop on "Photon and neutral meson physics at intermediate energies" (1987)

Chapter 9

9.1   G. Ramachandran, V. Devanathan: Nucl. Phys. **48**, 369 (1963)
9.2   G. Ramachandran, V. Devanathan: Nucl. Phys. **5**, 593 (1964)
9.3   V. Devanathan, R. Parthasarathy, P.R. Subramanian: Ann. Phys. (N.Y.) **73**, 291 (1972)
9.4   A. Possoz et al.: Phys. Lett. **50B**, 438 (1974); **70B**, 265 (1977);
       L.Ph. Roesch, V.L. Telegdi, P. Truttman, A. Zehnder, L. Grenacs, L. Palffy: Phys. Rev. Lett. **46**, 1507 (1981): Helv. Phys. Acta **55**, 74 (1982)
9.5   Y. Kuno, J. Imazato, K. Nishiyama, K. Nagamine, T. Yamazaki, T. Minamisono: Phys. Lett. **148B**, 270 (1984)
9.6   K. Shoda, H. Ohashi, K. Nakahara: Phys. Rev. Lett. **39**, 1131 (1977); Nucl. Phys. **A350**, 377 (1980)
9.7   K. Min, E.J. Winhold, K. Shoda, H. Tsubota, H. Ohashi, M. Yamazaki: Phys. Rev. Lett. **44**, 1384 (1980)
9.8   K. Min, E.J. Winhold, K. Shoda, M. Torikoshi, M. Yamazaki, O. Sasaki, H. Tsubota, B.N. Sung: Phys. Rev. **C28**, 464 (1983)
9.9   K. Shoda, M. Yamazaki, M. Torikoshi, H. Tsubota, K. Min, E.J. Winhold, A.M. Bernstein: Phys. Rev. **C27**, 443 (1983)
9.10   K. Shoda, M. Torikoshi, O. Sasaki, S. Toyama, T. Kobayashi, A. Kagaya, H. Tsubota: Phys. Rev. **C33**, 2179 (1986)
9.11   M. Yamazaki, K. Shoda, M. Torikoshi, O. Sasaki, H. Tsubota: Phys. Rev. **C34**, 1123 (1986)
9.12   K. Shoda, M. Torikoshi, M. Yamazaki, O. Sasaki, H. Tsubota: Nucl. Phys. **A439**, 669 (1985)
9.13   K. Shoda, A. Kagaya, S. Toyoma, M. Torikoshi, O. Sasaki, T. Kobayashi, H. Tsubota: Nucl. Phys. **A473**, 705 (1987)
9.14   N. Paras et al.: Phys. Rev. Lett. **42**, 1455 (1979)
9.15   R.H. Dalitz, D.R. Yennie: Phys. Rev. **105**, 1598 (1957)
9.16   P. Stoler et al.: Phys. Rev. **C22**, 911 (1980)
9.17   L. Tiator, D. Drechsel: Nucl. Phys. **A360**, 208 (1981)
9.18   L. Tiator, L.E. Wright: Nucl. Phys. **A379**, 407 (1982)
9.19   S. Furui: Nucl. Phys. **A300**, 385 (1978)
9.20   J. Asai, E.L. Tomusiak: Can. J. Phys. **55**, 2066 (1977); **56**, 1526 (E) (1978)
9.21   J.L. Sabutis, F. Tabakin: "Electroproduction of charged pions from light nuclei", Univ. of Pittsburgh, preprint
9.22   V. Devanathan: Nucl. Phys. **87**, 397 (1966)
9.23   V. Devanathan: Nucl. Phys. **87**, 256 (1966)

9.24 S. Fubini, Y. Nambu, V. Wataghin: Phys. Rev. **111**, 329 (1958)
9.25 G.G. Ohlsen: Phys. Rev. **120**, 584 (1960)
9.26 W. Czyz, J.D. Walecka: Nucl. Phys. **51**, 312 (1964)
9.27 E. Borie, H. Chandra, D. Drechsel: Nucl. Phys. **A226**, 58 (1974); Phys. Lett. **47B**, 292 (1973)
9.28 F. Borkowski et al.: Phys. Rev. Lett. **38**, 742 (1977)
9.29 T. Dressler, N. Freed, H.B. Miska: Phys. Rev. **C21**, 2119 (1980)
9.30 J. Cohen, J.M. Eisenberg: Phys. Rev. **C28**, 1309 (1983)
9.31 J. Asai, E.L. Tomusiak, E.T. Dressler: J. Phys. **G7**, 131 (1981)
9.32 D.M. Skopik, E.L. Tomusiak, J. Asai, J.J. Murphy: Phys. Rev. **C18**, 2219 (1978)
9.33 A.S. Rosenthal, D. Halderson, F. Tabakin: Phys. Lett. **182**, 143 (1986)
9.34 A.S. Rosenthal, D. Halderson, K. Hodgkinson, F. Tabakin: Ann Phys. (N.Y.) **184**, 33 (1988)
9.35 C. Mishra et al.: Bull. Amer. Phys. Soc. **33**, 902 (1988)
9.36 J.C. Peng et al.: Phys. Rev. Lett. **58**, 2027 (1987)
9.37 J.C. Peng: "η meson photoproduction experiments", in: Proc. of the LAMPF Workshop on Photon and Neutral Meson Physics at Intermediate Energies, 1986, ed. by H.W. Baer et al.
9.38 J. Berger et al.: Phys. Rev. Lett. **61**, 919 (1988)
9.39 L. Adiels et al.: π^0 and η Spectroscopy at LEAR, in: Physics with Antiprotons at LEAR via the ACOL Era, Tignes (1985)
9.40 F. Tabakin, S.A. Dytman, A.S. Rosenthal: "Phototproduction of η mesons", invited talk at the RPI Conference on Excited Baryons, Aug. 4–6, 1988
9.41 H.R. Hicks et al.: Phys. Rev. **D7**, 2614 (1973)
9.42 H.D. Holmgren: "A Window to the Future", in *Perspectives in Nuclear Physics*, ed. by V. Devanathan, P.R. Subramanian, University of Madras (1987)
9.43 D.W. Donnelly, S.R. Cotanch: Research Program at CEBAF, Report of the 1985 Summer Study Group
9.44 B.A. Mecking: Research Program at CEBAF, Report of the 1985 Summer Study Group
9.45 L.M. Nath, S.K. Singh: Phys. Rev. **C39**, 1207 (1989)
9.46 M. Yamazaki et al.: Phys. Rev. **C35**, 355 (1987)
9.47 R.A. Eramzhyan, M. Gmitro, S.S. Kamalov, R. Mach: J. Phys. **G9**, 605 (1983);
      R.A. Eramzhyan, M. Gmitro, S.S. Kamalov: Phys. Lett. **128B**, 371 (1983)
9.48 V. DeCarlo, N. Freed: Phys. Rev. **C25**, 2162 (1982)
9.49 V. Girija, V. Devanathan: Phys. Rev. **C26**, 2152 (1982)

Appendix

A.1 H. Überall: *Electron Scattering from Complex Nuclei* (Academic Press, New York, 1971)

Subject Index

Series Index

Springer Tracts in Modern Physics (STMP)
Volumes: STMP 36 – 120

This cumulative index is based upon the Physics and Astronomy Classification Scheme (PACS) developed by the American Institute of Physics. First authors are listed alphabetically under each of the PACS headings.

KENKRE V.M.
The Master Equation Approach: Co-
herence, Energy Transfer, Annihilation,
and Relaxation
STMP 94, 1–109 (1982)
REINEKER P.
Stochastic Liouville Equation Approach:
Coupled Coherent and Incoherent Mo-
tion, Optical Line Shapes, Magnetic Res-
onance Phenomena
STMP 94, 111–226 (1982)

07.65 Optical Spectroscopy and
Spectrometers

RAETHER H.
Surface Plasmons on Smooth and Rough
Surfaces and on Gratings
STMP 111, 1–136 (1988)

07.85 X – and γ–Ray Instruments
and Techniques

GODWIN R.P.
Synchrotron Radiation as a Light Source
STMP 51, 1–73 (1969)

10.00 Elementary Particle Physics

ATKINSON D.
Bifurcation Theory Applied to Chiral
Symmetry Breaking
STMP 119, 158–162 (1990)
BONNIER B., HONTEBEYRIE M.

On the Scaling Behaviour of the O(N)
σ-Model
STMP 119, 163–167 (1990)
BROS J., VIANO G.A.
Complex Angular Momentum Analysis
in Axiomatic Quantum Field Theory
STMP 119, 53–76 (1990)
CIULLI S., SCHECK F.,
THIRRING W. (Eds.)
Rigorous Methods in Particle Physics
STMP 119, 1–220 (1990)

GOUNARIS G., RENARD F.M., SCHILD-
KNECHT D.
Are Higgs Interactions Observable
at the 100 GeV Scale?
STMP 119, 188–201 (1990)
MENESSIER G., CAUSSE M.B., AUBER-
SON G.
A New Method for QCD Sum Rules
STMP 119, 137–157 (1990)
NARNHOFER H., THIRRING W.
A Model for a Dia-Electric
STMP 119, 1–3 (1990)
ROY S.M., SINGH V.
High Energy Theorems
STMP 119, 38–52 (1990)

SCHECK F.
Geometric Approaches to Particle Physics
STMP 119, 202–216 (1990)

11.00 General Theory of Fields and Particles

BRANDT R.A.
Physics on the Light Cone
STMP 57, 237–247 (1971)
FERRARA S., GATTO R., GRILLO A.F.
Conformal Algebra in Space-Time and
Operator Product Expansion
STMP 67, 1–64 (1973)
JACKIW R.
Canonical Light-Cone Commutators and
their Applications
STMP 62, 1–36 (1972)
KUNDT W.
Canonical Quantization of Gauge In-
variant Field Theories
STMP 40, 108–168 (1966)
RÜHL W.
Application of Harmonic Analysis to
Inelastic Electron-Proton Scattering
STMP 57, 202–221 (1971)
SYMANZIK K.
Small-Distance Behaviour in Field The-
ory
STMP 57, 222–236 (1971)
ZIMMERMANN W.
Problems in Vector Meson Theories
STMP 50, 143–156

11.30 Symmetry and Conservation Laws

BARUT A.O.
Dynamical Groups and their Currents.
A Model for Strong Interactions
STMP 50, 1–28 (1969)
EKSTEIN H.
Rigourous Symmetries of Elementary
Particles
STMP 37, 150–180 (1965)
LOPUSZANSKI J.T.
Physical Symmetries in the Framework
of Quantum Field Theory
STMP 52, 201–214 (1970)
PAULI W.
Continuous Groups in Quantum Me-
chanics
STMP 37, 85–104 (1965)
RACAH G.
Group Theory and Spectroscopy
STMP 37, 28–84 (1965)
RÜHL W.
Application of Harmonic Analysis to
Inelastic Electron-Proton Scattering
STMP 57, 202–221 (1971)
WESS J.
Realisations of a Compact, Connected,
Semisimple Lie Group
STMP 50, 132–142 (1969)

WESS J.
Conformal Invariance and the Energy-
Momentum Tensor
STMP 60, 1–17 (1971)

11.40 Currents and Their Properties

FURLAN G., PAVER N.,
VERZEGNASSI C.
Low Energy Theorems and Photo- and
Electroproduction Near Threshold by
Current Algebra
STMP 62, 118–147 (1972)
GATTO R.
Cabibbo Angle and SU2 × SU2 Break-
ing
STMP 53, 45–106 (1970)
GENZ H.
Local Properties of σ-Terms: A Review
STMP 61, 130–136 (1972)
KLEINER H.
Baryon Current Solving SU(3) Charge-
Current Algebra
STMP 49, 80–146 (1969)
LEUTWYLER H.
Current Algebra and Lightlike Charges
STMP 50, 29–41 (1969)
MENDES R.V., NE'EMAN Y.
Representations of the Local Current
Algebra
STMP 60, 18–31 (1971)
MÜLLER V.F.
Introduction to the Lagrangian Method
STMP 50, 42–52 (1969)
PIETSCHMANN H.
Introduction to the Method of Current
Algebra
STMP 50, 53–64 (1965)
PILKUHN H.
Coupling Constants from PCAC
STMP 55, 168–173 (1970)
PILKUHN H.
S-Matrix Formulation of Current Alge-
bra
STMP 50, 65–70 (1969)
RENNER B.
On the Problem of the Sigma Terms in
Meson-Baryon Scattering. Comments on
Recent Literature
STMP 61, 120–129 (1972)
RENNER B.
Current Algebra and Weak Interactions
STMP 52, 60–78 (1970)
SOLOVIEV L.D.
Symmetries and Current Algebras for
Electromagnetic Interactions
STMP 46, 53–66 (1968)
STECH B.
Nonleptonic Decays and Mass Differ-
ences of Hadrons
STMP 50, 84–99 (1969)

STICHEL P.
Introduction to Current Algebra
STMP 50, 120–131 (1969)
STICHEL P.
Current Algebra in the Framework of
General Quantum Field Theory
STMP 50, 100–109 (1969)
STICHEL P.
Current Algebra and Renormalizable
Field Theories
STMP 50, 110–119 (1969)
VERZEGNASSI C.
Low Energy Photo and Electroproduc-
tion, Multipole Analysis by Current Al-
gebra Commutators
STMP 59, 154–163 (1971)
WEINSTEIN M.
Chiral Symmetry. An Approach to the
Study of the Strong Interactions
STMP 60, 32–73 (1971)

12.00 Specific Theories and Interaction Models

AMALDI E., FUBINI S., FURLAN G.
Pion-Electroproduction. Electroproduc-
tion at Low Energy and Hadron Form
Factors
STMP 83, 1–162 (1979)
HOFMANN W.
Jets of Hadrons
STMP 90, 1–210 (1981)
KIESLING Ch.
Tests of the Standard Theory of Weak
Interactions
STMP 112, 1–212 (1988)
NAGL A., DEVANATHAN V., UEBER-
ALL H.
Nuclear Pion Photoproduction
STMP 90, 1–210 (1981)
WIIK B.H., WOLF G.
Electron-Positron Interactions
STMP 86, 1–262 (1979)

12.20 Quantum Electrodynamics

KÄLLEN G.
Radiative Corrections in Elementary Par-
ticle Physics
STMP 46, 67–132 (1968)
OLSEN H.A.
Applications of Quantum Electrodynam-
ics
STMP 44, 84–201 (1968)

12.30 Models of Weak Interactions

BARUT A.O.
On the S-Matrix Theory of Weak In-
teractions
STMP 53, 1–5 (1970)

HANAMURA E.
Biexcitons – Bose Condensation and Optical Response
STMP 73, 43–69 (1975)
KENKRE V.M.
The Master Equation Approach: Coherence, Energy Transfer, Annihilation, and Relaxation

STMP 94, 1–109 (1982)
LEVY R., BIVAS A., GRUN J., NIKITINE S.
Interaction between Excitons at High Concentration
STMP 73, 171–190 (1975)
MAHR H.
Medium and High Polariton Densities
STMP 73, 265–284 (1975)
NIKITINE S.
Properties of Biexcitons
STMP 73, 18–42 (1975)
NIKITINE S.
Introduction to Exciton Spectroscopy
STMP 73, 5–16 (1975)
NIMTZ G., SCHLICHT B.
Narrow-Gap Lead Salts
STMP 98, 1–117 (1983)

NOVIKOV B.V.
Spectroscopic Study of Exciton-Exciton Interaction (Biexcitons, Drops) in Semiconducting Crystals
STMP 73, 106–126 (1975)
PICK H.
Struktur von Störstellen in Alkalihalogenidkristallen
STMP 38, 2–83 (1965)
RAETHER H.
Solid State Excitations by Electrons. Plasm Oscillations and Single Electron Transitions
STMP 38, 85–157 (1965)
RASHBA E,I,
Gigantic Oscillator Strengths Inherent in Exciton Complexes
STMP 73, 150–170 (1975)
REINEKER P.
Stochastic Liouville Equation Approach: Coupled Coherent and Incoherent Motion, Optical Line Shapes, Magnetic Resonance Phenomena
STMP 94, 111–226 (1982)
RICE T.M.
Theory of Electron-Hole Drops in Germanium and Silicon
STMP 73, 91–194 (1975)
ROGACHEV A.A.
Exciton Condensation in Germanium
STMP 73, 129–148 (1975)

SCHMID D.
Nuclear Magnetic Double-Resonance – Principles and Applications in Solid-State Physics
STMP 68, 1–75 (1973)
SCHWENTNER N., KOCH E., JORTNER J.
Electronic Excitations in Condensed Rare Gases
STMP 107, 1–239 (1985)
SHAKLEE K.L.
Experimental Studies of Excitons at High Densities
STMP 73, 222–240 (1975)
STAHL A., BALSLEV I.
Electrodynamics of the Semiconductor Band Edge
STMP 110, 1–205 (1987)

72.00 Electronic Transport in Condensed Matter

BAUER G.
Determination of Electron Temperatures and of Hot Electron Distribution Functions in Semiconductors
STMP 74, 1–106 (1974)
DORNHAUS R., NIMTZ G.
The Properties and Applications of the Hg(1–x)Cd(x)Te Alloy System
STMP 98, 119–304 (1983)
FEITKNECHT J.
Silicon Carbide as a Semiconductor
STMP 58, 48–118 (1971)
GROSSE P.
Die Festkörpereigenschaften von Tellur
STMP 48, 1–208 (1969)
NIMTZ G., SCHLICHT B.
Narrow-Gap Lead Salts
STMP 98, 1–117 (1983)
SCHNAKENBERG J.
Electron-Phonon Interaction and the Boltzmann Equation in Narrow-Band Semiconductors
STMP 51, 74–120 (1969)
STAHL A., BALSLEV I.
Electrodynamics of the Semiconductor Band Edge
STMP 110, 1–205 (1987)

73.00 Properties of Surfaces and Thin Films

FORSTMANN F., GERHARDTS R.R.
Metal Optics Near the Plasma Frequency
STMP 109, 1–132 (1986)
HÖLZL J., SCHULTE F.K
Work Function of Metals
STMP 85, 1–150 (1979)
MÜLLER K.
How Much can Auger Electrons Tell us about Solid Surfaces?
STMP 77, 97–125 (1975)

RAETHER H.
Excitation of Plasmons and Interband
Transitions by Electrons
STMP 88, 1–196 (1980)
WAGNER H.
Physical and Chemical Properties of
Stepped Surfaces
STMP 85, 151–221 (1979)
WISSMANN P.
The Electrical Resistivity of Pure and
Gas Covered Metal Films
STMP 77, 1–96 (1975)

74.00 Superconductivity

LÜDERS G., USADEL K.-D.
The Method of the Correlation Func-
tion in Superconductivity Theory
STMP 56, 1–215 (1971)
ULLMAIER H.
Irreversible Properties of Type II Su-
perconductors
STMP 76, 1–165 (1975)

75.00 Magnetic Properties

FISCHER K.
Magnetic Impurities in Metals: The s-d
Exchange Model
STMP 54, 1–76 (1970)
REINEKER P.
Stochastic Liouville Equation Approach:
Coupled Coherent and Incoherent Mo-
tion, Optical Line Shapes, Magnetic Res-
onance Phenomena
STMP 94, 111–226 (1982)
STIERSTADT K.
Der Magnetische Barkhausen-Effekt
STMP 40, 3–106 (1966)

76.00 Magnetic Resonances and Relaxation

COUFAL H., LÜSCHER E.,
MICKLITZ H.
Electron Spin and Nuclear Gamma Res-
onance Studies of Rare Gas Matrix-
Isolated Atoms and Ions
STMP 103, 1–57 (1984)
NORBERG R.E.
Nuclear Magnetic Resonance in Con-
densed Rare Gases
STMP 103, 59–95 (1984)
SCHMID D.
Nuclear Magnetic Double-Resonance –
Principles and Applications in Solid-
State Physics
STMP 68, 1–75 (1973)

77.00 Dielectric Properties and Materials

BUSSMANN-HOLDER, BILZ H.,
VOGT P.
Electronic and Dynamical Properties of
IV – IV Compounds
STMP 99, 51–98 (1983)

JANTSCH W.
Dielectric Properties and Soft Modes in
Semiconducting (Pb, Sn, Ge)Te
STMP 99, 1–50 (1983)

78.00 Optical Properties

BÄUERLE D.
Vibrational Spectra of Electron and Hy-
drogen Centers in Ionic Crystals
STMP 68, 76–160 (1973)
BORSTEL G., FALGE H.J., OTTO A.
Surface and Bulk Phonon-Polaritons Ob-
served by Attenuated Total Reflection
STMP 74, 107–148 (1974)

BUSSMANN-HOLDER, BILZ H.,
VOGT P.
Electronic and Dynamical Properties of
IV – IV Compounds
STMP 99, 51–98 (1983)
CLAUS R., MERTEN L.,
BRANDMÜLLER J.
Light Scattering by Phonon-Polaritons
STMP 75, 1–237 (1975)
DANIELS J., FESTENBERG C.v.,
RAETHER H., ZEPPENFELD K.
Optical Constants of Solids by Electron
Spectroscopy
STMP 54, 77–135 (1970)

DORNHAUS R., NIMTZ G.
The Properties and Applications of the
Hg(1–x)Cd(x)Te Alloy System
STMP 98, 119–304 (1983)
FORSTMANN F., GERHARDTS R.R.
Metal Optics Near the Plasma Frequency
STMP 109, 1–132 (1986)
GODWIN R.P.
Synchrotron Radiation as a Light Source
STMP 51, 1–73 (1969)
GROSSMANN M., BIELMANN J., NIKI-
TINE S.
Tests of Validity of Spatial Dispersion
Theories on Lead Iodide Crystal Spec-
tra
STMP 73, 243–264 (1975)
HAKEN H., NIKITINE S. (Eds.)
Excitons at High Densities
STMP 73, 1–298 (1975)

HAKEN H., NIKITINE S.
Theory of Stimulated Emission by Ex-
citons
STMP 73, 192–210 (1975)
JANTSCH W.
Dielectric Properties and Soft Modes in
Semiconducting (Pb, Sn, Ge)Te
STMP 99, 1–50 (1983)

LENGELER B.
De Haas – van Alphen Studies of the
Electronic Structure of the Noble Metals and Their Dilute Alloys
STMP *82*, 1–67 (1978)
LEVY R., BIVAS A., GRUN J.B., NIKITINE S.
Interaction between Excitons at High
Concentration
STMP *73*, 171–190 (1975)
LEVY R., GRUN J.B., NIKITINE S.
Experimental Investigation of the Competition of Stimulated Emission Involving Exiton
STMP *73*, 211–219 (1975)
NIMTZ G., SCHLICHT B.
Narrow-Gap Lead Salts
STMP *98*, 1–117 (1983)
POCKRAND I.
Surface Enhanced Raman Vibrational
Studies at Solid/Gas Interfaces
STMP *104*, 1–164 (1984)
RAETHER H.
Solid State Excitations by Electrons.
Plasma Oscillations and Single Electron
Transitions
STMP *38*, 85–157 (1965)

RICHTER W.
Resonant Raman Scattering in Semiconductors
STMP *78*, 121–272 (1976)
STAHL A., BALSLEV I.
Electrodynamics of the Semiconductor
Band Edge
STMP *110*, 1–205 (1987)

79.00 Electron and Ion Emission by Liquids
and Solids; Impact Phenomena
KIRSCHNER J.
Polarized Electrons at Surfaces
STMP *106*, 1–158 (1985)

80.00 CROSS-DISCIPLINARY PHYSICS
AND RELATED AREAS OF SCIENCE
AND TECHNOLOGY

81.00 Material Sciences
KHABIBULLAEV P.K.,
SKORODUMOV B.G.
Determination of Hydrogen in Materials – Nuclear Phycics Methods
STMP *117*, 1–87 (1989)
OVERHOF H., THOMAS P.
Electronic Transport in Hydrogenated
Amorphous Semiconductors
STMP *114*, 1–174 (1989)

POELSEMA B., COMSA G.
Scattering of Thermal Energy Atoms
from Disordered Surfaces
STMP *115*, 1–108 (1989)

82.00 Physical Chemistry
LECHNER R.E., RIEKEL C.
Applications of Neutron Scattering in
Chemistry
STMP *101*, 1–84 (1983)

85.70 Magnetic Devices
LEHNER G.
Über die Grenzen der Erzeugung sehr
hoher Magnetfelder
STMP *47*, 67–110 (1968)

90.00 GEOPHYSICS, ASTRONOMY
AND ASTROPHYSICS

95.00 Theoretical Astrophysics
KUNDT W.
Survey of Cosmology. Is "Our World"
Implied by Thermal Equilibrium in the
Hadron Era?
STMP *58*, 1–47 (1971)
KUNDT W.
Recent Progress in Cosmology. Isotropy
of the 3 deg Background Radiation, and
the Occurrence of Spacetime Singularities
STMP *47*, 111–142 (1968)
STEWART J., WALKER M.
Black Holes: the Outside Story
STMP *69*, 69–115 (1973)

97.00 Stars
BÖRNER G.
On the Properties of Matter in Neutron
Stars
STMP *69*, 1–67 (1973)

Author Index

Springer Tracts in Modern Physics (STMP)
Volumes: STMP 36 – 120

Alphabetical list of all contributors together with name of first author,
volume number, page numbers, year of publication and main PACS number.

Name	First Author	STMP	Pages			Year	PACS
Ademollo M.	Ademollo M.	59	135	–	153	(71)	12.40
Agarwal G.S.	Agarwal G.S.	70	1	–	129	(74)	42.50
Amaldi E.	Amaldi E.	83	1	–	162	(79)	12.00
Arenhövel H.	Arenhövel H.	65	58	–	91	(72)	21.00
Atkinson D.	Atkinson D.	119	158	–	162	(90)	10.00
Atkinson D.	Atkinson D.	57	1	–	21	(71)	13.75
Auberson G.	Auberson G.	119	26	–	37	(90)	03.98
Auberson G.	Mennessier G.	119	137	–	157	(90)	10.00
Bagaev V.S.	Bagaev V.S.	73	72	–	90	(75)	71.00
Balslev I.	Stahl A.	110	1	–	205	(87)	71.35
Baltay C.	Baltay C.	108	7	–	15	(86)	29.00
Barut A.O.	Barut A.O.	50	1	–	28	(69)	11.30
Barut A.O.	Barut A.O.	53	1	–	5	(70)	12.30
Basdevant J.L.	Basdevant J.L.	61	1	–	24	(72)	13.75
Bauer G.	Bauer G.	74	1	–	106	(74)	72.00
Bäuerle D.	Bäuerle D.	68	76	–	160	(73)	78.00
Baym G.	Baym G.	100	186	–	213	(82)	21.65
Behringer J.	Behringer J.	68	161	–	199	(73)	61.00
Bendow B.	Bendow B.	82	69	–	114	(78)	71.00
Bennemann K.H.	Bennemann K.H.	38	158	–	188	(65)	71.00
Bielmann J.	Grossmann M.	73	243	–	264	(75)	78.20
Bilz H.	Bussmann-Holder	99	51	–	98	(83)	77.00
Bivas A.	Levy R.	73	171	–	190	(75)	71.00
Bjorken J.D.	Bjorken J.D.	108	17	–	30	(86)	29.00
Blum W.	Blum W.	108	31	–	40	(86)	29.00
Börner G.	Börner G.	69	1	–	67	(73)	97.00
Bonnier B.	Bonnier B.	119	163	–	167	(90)	10.00
Borstel G.	Borstel G.	74	107	–	148	(74)	78.00
Brandmüller J.	Claus R.	75	1	–	237	(75)	78.00
Brandt R.A.	Brandt R.A.	57	237	–	247	(71)	11.00
Breuer N.	Leibfried G.	81	1	–	342	(78)	61.00
Brinckmann P.	Brinckmann P.	61	135	–	166	(72)	13.60
Brodsky S.J.	Brodsky S.J.	100	81	–	144	(82)	12.35
Bros J.	Bros J.	119	53	–	76	(90)	03.00

Name	First Author	STMP	Pages			Year	PACS
Buchanan C.D.	Buchanan C.D.	*39*	20	–	42	(65)	13.40
Bussmann-Holder	Bussmann-Holder	*99*	51	–	98	(83)	77.00
Büttgenbach S.	Büttgenbach S.	*96*	1	–	97	(82)	32.00
Cannata F.	Cannata F.	*89*	1	–	112	(80)	21.00
Causse M.B.	Mennessier G.	*119*	137	–	157	(90)	10.00
Chadan K.	Chadan K.	*119*	14	–	19	(90)	03.98
Charpak G.	Charpak G.	*108*	41	–	48	(80)	29.00
Ciulli M.	Ciulli M.	*119*	102	–	125	(90)	02.98
Ciulli S.	Ciulli M.	*119*	102	–	125	(90)	02.98
Ciulli S.	Ciulli S.	*119*	1	–	220	(90)	10.00
Claus R.	Claus R.	*75*	1	–	237	(75)	78.00
Close F.E.	Close F.E.	*100*	57	–	80	(82)	12.35
Collard H.	Buchanan C.D.	*39*	10	–	42	(65)	13.40
Collins P.D.B.	Collins P.D.B.	*45*	1	–	292	(68)	12.40
Collins P.D.B.	Collins P.D.B.	*60*	204	–	233	(71)	12.40
Collins P.D.B.	Collins P.D.B.	*63*	163	–	189	(72)	12.40
Common A.K.	Common A.K.	*119*	20	–	25	(90)	03.00
Comsa G.	Poelsema B.	*115*	1	–	108	(89)	81.00
Contogouris A.P.	Contogouris A.P.	*57*	92	–	118	(71)	12.40
Contogouris A.P.	Contogouris A.P.	*63*	145	–	162	(72)	12.40
Cornille H.	Cornille H.	*119*	168	–	187	(90)	02.20
Coufal H.	Coufal H.	*103*	1	–	57	(84)	76.80
Crannell C.	Buchanan C.D.	*39*	20	–	42	(65)	13.40
Daniels J.	Daniels J.	*54*	77	–	135	(70)	71.00
Dederichs P.H.	Dederichs P.H.	*87*	1	–	170	(80)	61.00
Dettmann K.	Dettmann K.	*58*	119	–	206	(71)	34.00
Devanathan V.	Nagl A.	*120*	1	–	192	(91)	12.00
Dietz K.	Dietz K.	*60*	74	–	90	(71)	12.40
Donnachie A.	Donnachie A.	*61*	25	–	48	(72)	14.20
Donnachie A.	Donnachie A.	*63*	121	–	144	(72)	13.60
Donner W.	Donner W.	*37*	1	–	27	(65)	31.00
Dorner B.	Dorner B.	*93*	1	–	96	(82)	61.12
Dornhaus R.	Dornhaus R.	*78*	1	–	120	(76)	71.00
Dornhaus R.	Dornhaus R.	*98*	119	–	304	(83)	71.00
Dosch H.G.	Dosch H.G.	*52*	79	–	90	(70)	12.30
Drees J.	Drees J.	*60*	107	–	137	(71)	13.60
Drell S.D.	Drell S.D.	*39*	71	–	90	(65)	13.60
Drell S.D.	Drell S.D.	*108*	49	–	69	(86)	29.00
Dydak F.	Dydak F.	*108*	71	–	93	(86)	29.00
Ebel G.	Ebel G.	*55*	239	–	290	(70)	13.75
Ehrfeld W.	Ehrfeld W.	*97*	1	–	140	(83)	47.00
Eisele F.	Eisele F.	*108*	95	–	111	(86)	29.00
Ekstein H.	Ekstein H.	*37*	150	–	180	(65)	11.30
Ellis J.	Ellis J.	*108*	113	–	122	(86)	29.00
Engel T.	Engel T.	*91*	55	–	180	(82)	68.00
Faessler A.	Faessler A.	*100*	214	–	223	(82)	21.00
Falge H.J.	Borstel G.	*74*	107	–	148	(74)	78.00
Feitknecht J.	Feitknecht J.	*58*	48	–	118	(71)	72.00
Ferrara S.	Ferrara S.	*67*	1	–	64	(73)	11.00
Festenberg C.v.	Daniels J.	*54*	77	–	135	(70)	71.00
Fischer H.	Fischer H.	*59*	188	–	222	(71)	13.60

168

Name	First Author	STMP	Pages			Year	PACS
Fischer K.	Fischer K.	54	1	–	76	(70)	75.00
Flügge G.	Flügge G.	100	1	–	55	(82)	13.60
Foa L.	Foa L.	59	114	–	134	(71)	13.60
Forstmann F.	Forstmann F.	109	1	–	132	(86)	71.45
Fries D.	Fries D.	100	1	–	223	(82)	21.00
Frosch R.	Buchanan C.D.	39	20	–	42	(65)	13.40
Froyland J.	Froyland J.	63	1	–	30	(72)	13.60
Fubini S.	Amaldi E.	83	1	–	162	(79)	12.00
Furlan G.	Furlan G.	62	118	–	147	(72)	13.60
Furlan G.	Amaldi E.	83	1	–	162	(79)	12.00
Gaillard M.K.	Gaillard M.K.	108	123	–	138	(86)	29.00
Gasiorowicz S.	Gasiorowicz S.	52	1	–	33	(70)	12.30
Gatto R.	Gatto R.	39	106	–	137	(65)	13.40
Gatto R.	Gatto R.	53	45	–	106	(70)	12.30
Gatto R.	Gatto R.	67	1	–	64	(73)	11.00
Gault F.D.	Collins P.D.B.	63	163	–	189	(72)	12.40
Gehlen G.von	Gehlen G.von	53	29	–	44	(70)	12.30
Gehlen G.von	Gehlen G.von	59	164	–	187	(71)	13.60
Geiger W.	Geiger W.	46	1	–	52	(68)	51.00
Genz H.	Genz H.	61	130	–	136	(72)	11.40
Gerhardts R.R.	Forstmann F.	109	1	–	132	(86)	71.45
Geweniger C.	Dydak F.	108	71	–	93	(86)	29.00
Godwin R.P.	Godwin R.P.	51	1	–	73	(69)	42.72
Goll J.	Haken H.	73	285	–	295	(75)	71.00
Gounaris G.	Gounaris G.	119	188	–	201	(90)	10.00
Gourdin M.	Gourdin M.	55	192	–	212	(70)	13.40
Grabert H.	Grabert H.	95	1	–	164	(82)	05.00
Graham R.	Graham R.	66	1	–	97	(73)	42.50
Grillo A.F.	Ferrara S.	67	1	–	64	(73)	11.00
Grosse P.	Grosse P.	48	1	–	208	(69)	71.00
Grosse H.	Grosse H.	119	4	–	13	(90)	03.00
Grossmann M.	Grossmann M.	73	243	–	264	(75)	78.20
Grun J.B.	Levy R.	73	171	–	190	(75)	71.00
Grun J.B.	Levy R.	73	211	–	219	(75)	78.00
Gustafson G.	Gustafson G.	61	49	–	67	(72)	13.75
Haake F.	Haake F.	66	98	–	168	(73)	42.50
Haken H.	Haken H.(ed)	73	1	–	298	(75)	71.00
Haken H.	Haken H.	73	192	–	210	(75)	71.00
Haken H.	Haken H.	73	285	–	295	(75)	71.00
Hamilton J.	Hamilton J.	57	41	–	70	(71)	13.75
Hamilton J.	Gustafson G.	61	49	–	67	(72)	13.75
Hanamura E.	Hanamura E.	73	43	–	69	(75)	71.00
Hasegawa A.	Hasegawa A.	116	1	–	75	(89)	42.50
Hawkes P.W.	Hawkes P.W.	42	1	–	126	(66)	41.80
Heinloth K.	Heinloth K.	65	92	–	145	(72)	13.60
Heintzmann H.	Heintzmann H.	47	185	–	225	(68)	04.00
Heinz K.	Heinz K.	91	1	–	53	(82)	68.00
Heller L.	Heller L.	100	145	–	185	(82)	12.35
Henley E.M.	Henley E.M.	108	139	–	147	(86)	29.00
Hess S.	Hess S.	54	136	–	176	(70)	51.00
Hofmann W.	Hofmann W.	90	1	–	210	(81)	12.00

Name	First Author	STMP	Pages			Year	PACS
Hofstadter R.	Buchanan C.D.	*39*	20	–	42	(65)	13.40
Höhler G.	Höhler G.	*39*	55	–	70	(65)	13.60
Holtey G.von	Holtey G.von	*59*	3	–	26	(71)	13.60
Hölzl J.	Hölzl J.	*85*	1	–	150	(79)	73.00
Hontebeyrie M.	Bonnier B.	*119*	163	–	167	(90)	10.00
Hornberg H.	Geiger W.	*46*	1	–	52	(68)	51.00
Hove L.van	Hove L.van	*39*	1	–	19	(65)	12.40
Huang K.	Huang K.	*62*	98	–	106	(72)	13.40
Huang K.	Huang K.	*62*	107	–	117	(72)	12.40
Jackiw R.	Jackiw R.	*62*	1	–	36	(72)	11.00
Jantsch W.	Jantsch W.	*99*	1	–	50	(83)	77.00
Jortner J.	Schwentner N.	*107*	1	–	239	(85)	71.00
Julius D.	Ebel G.	*55*	239	–	290	(70)	13.75
Kabir P.K.	Kabir P.K.	*52*	91	–	112	(70)	12.30
Källen G.	Källen G.	*46*	67	–	132	(68)	12.20
Kenkre V.M	Kenkre V.M.	*94*	1	–	109	(82)	05.00
Khabibullaev P.K.	Khabibullaev P.K.	*117*	1	–	87	(89)	81.00
Khuri N.N.	Khuri N.N.	*119*	77	–	97	(90)	02.98
Kiesling Ch.	Kiesling Ch.	*112*	1	–	212	(88)	12.00
Kirschner J.	Kirschner J.	*106*	1	–	158	(85)	79.20
Kleiner H.	Kleiner H.	*49*	90	–	146	(69)	11.40
Kleinknecht K.	Kleinknecht K.	*108*	1	–	291	(86)	29.00
Kleinknecht K.	Kleinknecht K.	*108*	149	–	164	(86)	29.00
Kobayashi R.	Chadan K.	*119*	14	–	19	(90)	03.98
Koch E.-E.	Schwentner N.	*107*	1	–	239	(85)	71.00
Koester L.	Koester L.	*80*	1	–	55	(77)	28.20
Kolanoski H.	Kolanoski H.	*105*	1	–	187	(84)	13.65
Kramer G.	Kramer G.	*55*	152	–	167	(70)	13.75
Kramer G.	Kramer G.	*102*	1	–	140	(84)	13.65
Kummer W.	Kummer W.	*52*	113	–	125	(70)	12.30
Kundt W.	Kundt W.	*40*	108	–	168	(66)	11.00
Kundt W.	Kundt W.	*47*	111	–	142	(68)	95.00
Kundt W.	Kundt W.	*58*	1	–	47	(71)	95.00
Kuyucak S.	Henley E.M.	*108*	139	–	147	(86)	29.00
Lacmann R.	Lacmann R.	*44*	1	–	81	(68)	61.00
Landshoff P.V.	Landshoff P.V.	*62*	37	–	50	(72)	13.60
Langbein D.	Langbein D.	*72*	1	–	139	(74)	34.00
Lechner R.E.	Lechner R.E.	*101*	1	–	84	(83)	61.21
Lederman L.M	Lederman L.M.	*108*	165	–	184	(86)	29.00
Lee T.D.	Kleinknecht K.	*108*	1	–	291	(86)	29.00
Lee T.D.	Lee T.D.	*108*	185	–	189	(86)	29.00
Lehner G.	Lehner G.	*47*	67	–	110	(68)	85.70
Leibfried G.	Leibfried G.	*81*	1	–	342	(78)	61.00
Lengeler B	Lengeler B.	*82*	1	–	67	(78)	78.00
Leon J.	Leon J.	*119*	98	–	101	(90)	03.98
Leutwyler H.	Leutwyler H.	*50*	29	–	41	(69)	11.40
Levinger J.S.	Levinger J.S.	*71*	88	-	240	(74)	21.40
Levy R.	Levy R.	*73*	171	–	190	(75)	71.00
Levy R.	Levy R.	*73*	211	–	219	(75)	78.00
Li M.	Henley E.M.	*108*	139	–	147	(86)	29.00
Lopuszanski J.T.	Lopuszanski J.T.	*52*	201	–	214	(70)	11.30

Name	First Author	STMP	Pages			Year	PACS
Lüders G.	Lüders G.	56	1	–	215	(71)	74.00
Ludwig W.	Ludwig W.	43	1	–	299	(66)	63.00
Lüke D.	Lüke D.	59	39	–	76	(71)	13.60
Lüscher E.	Coufal H.	103	1	–	57	(84)	76.80
Mahr H.	Mahr H.	73	265	–	284	(75)	71.00
Martin A.D.	Martin A.D.	55	142	–	151	(70)	13.75
Martin B.R.	Martin B.R.	55	74	–	141	(70)	13.75
May J.	Blum W.	108	31	–	40	(86)	29.00
McClure W.	Wildermuth K.	41	1	–	172	(66)	21.00
Mendes R.V.	Mendes R.V.	60	18	–	31	(71)	11.40
Mennessier G.	Mennessier G.	119	137	–	157	(90)	10.00
Merten L.	Claus R.	75	1	–	237	(75)	78.00
Michael C.	Michael C.	55	174	–	191	(70)	12.40
Micklitz H.	Coufal H.	103	1	–	57	(84)	76.80
Mittelstaedt P.	Heintzmann H.	47	185	–	225	(68)	04.00
Morgan D.	Morgan D.	55	1	–	42	(70)	13.75
Müllensiefen A.	Ebel G.	55	239	–	290	(70)	13.75
Müller K.	Müller K.	77	97	–	125	(75)	73.00
Müller K.	Heinz K.	91	1	–	53	(82)	68.00
Müller V.F.	Müller V.F.	50	42	–	52	(69)	11.40
Müller V.F.	Müller V.F.	52	34	–	49	(70)	12.30
Nagels M.M.	Swart J.J.de	60	138	–	203	(71)	13.75
Nagl A.	Nagl A.	120	1	–	192	(91)	12.00
Narnhofer H.	Narnhofer H.	119	1	–	3	(91)	10.00
Nauenberg U.	Nauenberg U.	108	191	–	223	(86)	29.00
Ne'eman Y.	Mendes R.V.	60	18	–	31	(71)	11.40
Neveu A.	Neveu A.	119	126	–	136	(90)	02.20
Nikitine S.	Nikitine S.	73	5	–	16	(75)	71.00
Nikitine S.	Nikitine S.	73	18	–	42	(75)	71.00
Nikitine S.	Levy R.	73	171	–	190	(75)	71.00
Nikitine S.	Haken H.	73	192	–	210	(75)	71.00
Nikitine S.	Levy R.	73	211	–	219	(75)	78.00
Nikitine S.	Grossmann M.	73	243	–	264	(75)	78.20
Nimtz G.	Dornhaus R.	78	1	–	120	(76)	71.00
Nimtz G.	Nimtz G.	98	1	–	117	(83)	71.00
Nimtz G.	Dornhaus R.	98	119	–	304	(83)	71.00
Norberg R.E.	Norberg R.E.	103	59	–	95	(84)	76.60
Novikov B.V.	Novikov B.V.	73	106	–	126	(75)	71.00
Oades G.C.	Oades G.C.	55	61	–	72	(70)	13.75
Oehme R.	Oehme R.	57	119	–	131	(71)	12.40
Oehme R.	Oehme R.	57	132	–	157	(71)	12.40
Oehme R.	Oehme R.	61	109	–	119	(72)	12.40
Olson C.L.	Olson C.L.	84	1	–	144	(79)	41.70
Olsen H.A.	Olsen H.A.	44	84	–	201	(68)	12.20
Osborne L.S.	Osborne L.S.	39	91	–	105	(65)	13.60
Otto A.	Borstel G.	74	107	–	148	(74)	78.00
Overhof H.	Overhof H.	114	1	–	174	(89)	81.00
Panofsky W.K.H.	Panofsky W.K.H.	39	138	–	154	(65)	29.00
Panofsky W.K.H.	Panofsky W.K.H.	108	225	–	235	(86)	29.00
Paul E.	Paul E.	79	53	–	145	(76)	12.30

Name	First Author	STMP	Pages			Year	PACS
Pauli W.	Pauli W.	*37*	85	–	104	(65)	11.30
Paver N.	Furlan G.	*62*	118	–	147	(72)	13.60
Pfeil W.	Pfeil W.	*55*	313	–	137	(70)	13.60
Pick H.	Pick H.	*38*	2	–	83	(65)	71.00
Pietschmann H.	Pietschmann H.	*50*	53	–	64	(69)	11.40
Pietschmann H.	Pietschmann H.	*52*	193	–	200	(70)	12.30
Pilkuhn H.	Pilkuhn H.	*50*	65	–	70	(69)	11.40
Pilkuhn H.	Pilkuhn H.	*55*	168	–	173	(70)	11.40
Pilkuhn H.	Ebel G.	*55*	239	–	290	(70)	13.75
Pisut J.	Morgan D.	*55*	239	–	290	(70)	13.75
Pisut J.	Pisut J.	*55*	43	–	60	(70)	13.75
Pockrand I.	Pockrand I.	*104*	1	–	164	(84)	78.30
Poelsema B.	Poelsema B.	*115*	1	–	108	(89)	81.00
Press W.	Press W.	*92*	1	–	129	(81)	61.12
Primakoff H.	Primakoff H.	*53*	6	–	28	(70)	12.30
Putlitz G.zu	Putlitz G.zu	*37*	106	–	149	(65)	21.10
Racah G.	Racah G.	*37*	28	–	84	(65)	21.00
Raether H.	Raether H.	*38*	85	–	157	(65)	71.00
Raether H.	Daniels J.	*54*	77	–	135	(70)	71.00
Raether H.	Raether H.	*88*	1	–	196	(80)	73.00
Raether H.	Raether H.	*111*	1	–	136	(88)	07.65
Rashba E.I.	Rashba E.I.	*73*	150	–	170	(75)	71.00
Ravenhall D.G.	Buchanan C.D.	*39*	20	–	42	(65)	13.40
Reineker P.	Reineker P.	*94*	111	–	226	(82)	05.00
Reiss H.	Reiss H.	*113*	1	–	201	(89)	42.10
Renard F.M.	Renard F.M.	*63*	98	–	120	(72)	13.60
Renard F.M.	Gounaris G.	*119*	188	–	201	(90)	10.00
Renner B.	Renner B.	*52*	60	–	78	(70)	11.40
Renner B.	Renner B.	*61*	120	–	129	(72)	11.40
Riazzuddin	Riazzuddin	*52*	126	–	160	(70)	12.30
Rice T.M.	Rice T.M.	*73*	91	–	104	(75)	71.00
Richter D.	Richter d.	*101*	85	–	222	(83)	66.30
Richter W.	Richter W.	*78*	121	–	272	(76)	78.00
Rieder K.-H.	Engel T.	*91*	55	–	180	(82)	68.00
Riekel C.	Lechner R.E.	*101*	1	–	84	(83)	61.12
Rijken T.A.	Swart J.J.de	*60*	138	–	203	(71)	13.75
Rittenberg V.	Rittenberg V.	*62*	92	–	97	(72)	13.60
Robrock K.-H.	Robrock K.- H.	*118*	1	–	106	(90)	46.00
Rogachev A.A.	Rogachev A.A.	*73*	129	–	148	(75)	71.00
Rolandi L.	Blum W.	*108*	31	–	40	(86)	29.00
Rollnik H.	Rollnik H.	*79*	1	–	52	(76)	13.60
Rothleitner J.	Rothleitner J.	*50*	71	–	83	(69)	14.20
Rothleitner J.	Rothleitner J.	*52*	161	–	170	(70)	12.30
Roy S.M.	Roy S.M.	*119*	38	–	52	(90)	10.00
Rubbia C.	Rubbia C.	*108*	237	–	266	(86)	29.00
Rubinstein H.R.	Rubinstein H.R.	*57*	193	–	201	(71)	12.40
Rubinstein H.R.	Rubinstein H.R.	*62*	72	–	91	(72)	12.40
Rühl W.	Rühl W.	*57*	202	–	221	(71)	13.60
Sabatier P.C.	Leon J.	*119*	98	–	101	(90)	03.98
Samios N.P.	Samios N.P.	*108*	267	–	268	(86)	29.00
Satz H.	Satz H.	*57*	158	–	190	(71)	12.40

Name	First Author	STMP	Pages			Year	PACS
Scheck F.	Ciulli S.	*119*	1	–	220	90)	10.00
Scheck F.	Scheck F.	*119*	202	–	216	(90)	10.00
Schenzle A.	Haken H.	*73*	285	–	295	(75)	71.00
Schildknecht D.	Schildknecht D.	*63*	57	–	97	(72)	13.60
Schildknecht D.	Gounaris G.	*119*	188	–	201	(90)	10.00
Schilling K.	Schilling K.	*63*	31	–	56	(72)	12.60
Schlicht B.	Nimtz G.	*98*	1	–	117	(83)	71.00
Schmid D.	Schmid D.	*68*	1	–	75	(73)	76.70
Schmidt W.	Ebel G.	*55*	239	–	290	(70)	13.75
Schnakenberg J.	Schnakenberg J.	*51*	74	–	120	(69)	72.00
Schramm K.H.	Geiger W.	*46*	1	–	52	(68)	51.00
Schramm K.H.	Schramm K.H.	*58*	207	–	265	(71)	63.00
Schrempp-Otto B.	Schrempp-Otto B.	*61*	68	–	108	(72)	12.40
Schrempp F.	Schrempp-Otto B.	*61*	68	–	108	(72)	12.40
Schroeder K.	Schroeder K.	*87*	171	–	262	(80)	61.00
Schulte F.K.	Hölzl J.	*85*	1	–	150	(79)	73.00
Schumacher U.	Schumacher U.	*84*	145	–	231	(79)	41.70
Schwartz M	Schwartz M.	*108*	269	–	274	(86)	29.00
Schwela D.	Pfeil W.	*55*	213	–	237	(70)	13.60
Schwela D.	Schwela D.	*59*	27	–	38	(71)	13.60
Schwentner N.	Schwentner N.	*107*	1	–	239	(85)	71.00
Segre G.	Segre G.	*52*	171	–	192	(70)	12.30
Seiwert R.	Seiwert R.	*47*	143	–	184	(68)	34.00
Shaklee K.L.	Shaklee K.L.	*73*	222	–	240	(75)	71.00
Singer P.	Singer P.	*71*	39	–	87	(74)	21.00
Singh V.	Roy S.M.	*119*	38	–	52	(90)	10.00
Skorodumov B.G.	Khabibullaev P.K.	*117*	1	–	87	(89)	81.00
Smith C.H.L.	Smith C.H.L.	*62*	51	–	71	(72)	13.60
Söding P.	Lüke D.	*59*	39	–	76	(71)	13.60
Soloviev L.D.	Soloviev L.D.	*46*	53	–	66	(68)	11.40
Spearman T.D.	Ciulli M.	*119*	102	–	125	(90)	02.98
Springer T.	Springer T.	*64*	1	–	100	(72)	28.20
Squires E.J.	Collins P.D.B.	*45*	1	–	292	(68)	12.40
Squires E.J.	Squires E.J.	*57*	71	–	91	(71)	12.40
Stahl A.	Stahl A.	*110*	1	–	205	(87)	71.35
Stech B.	Stech B.	*50*	84	–	99	(69)	11.40
Stech B.	Stech B.	*52*	50	–	59	(70)	12.30
Steeb S.	Steeb S.	*47*	2	–	66	(68)	61.00
Stewart J.	Stewart J.	*69*	69	–	115	(73)	97.60
Steyerl A.	Steyerl A.	*80*	57	–	130	(77)	28.20
Stichel P.	Stichel P.	*50*	100	–	109	(69)	11.40
Stichel P.	Stichel P.	*50*	110	–	119	(69)	11.40
Stichel P.	Stichel P.	*50*	120	–	131	(69)	11.40
Stichel P.	Rollnik H.	*79*	1	–	52	(76)	13.60
Stierstadt K.	Stierstadt K.	*40*	3	–	106	(66)	75.00
Strauch K.	Strauch K.	*39*	155	–	168	(65)	29.00
Su R.	Henley E.M.	*108*	139	–	147	(86)	29.00
Süßmann G.	Donner W.	*37*	1	–	27	(65)	31.00
Swart J.J. de	Ebel G.	*55*	239	–	290	(70)	13.75
Swart J.J.de	Swart J.J.de	*60*	138	–	203	(71)	13.75
Symanzik K.	Symanzik K.	*57*	222	–	236	(71)	11.00

Name	First Author	STMP	Pages			Year	PACS
Tan C.-I.	Tan C.-I.	60	91	–	106	(71)	12.40
Theissen H.	Theissen H.	65	91	–	106	(71)	25.30
Thirring W.	Ciulli S.	119	1	–	220	90)	10.00
Thirring W.	Narnhofer H.	119	1	–	3	(90)	10.00
Thomas P.	Overhof H.	114	1	–	174	(89)	81.00
Turlay R.	Eisele F.	108	95	–	111	(86)	29.00
Überall H.	Überall H.	49	1	–	89	(69)	25.30
Überall H.	Überall H.	71	1	–	38	(74)	21.00
Überall H.	Cannata F.	89	1	–	112	(80)	21.00
Überall H.	Nagle A.	120	1	–	192	(91)	12.00
Ullmaier H.	Ullmaier H.	76	1	–	165	(75)	74.00
Usadel K.-D.	Lüders G.	56	1	–	215	(71)	74.00
Verhoeven P.A.	Swart J.J.de	60	138	–	203	(71)	13.75
Verzegnassi C.	Verzegnassi C.	59	154	–	163	(71)	11.40
Verzegnassi C.	Furlan G.	62	118	–	147	(72)	13.60
Viano G.A.	Bros J.	119	53	–	76	(90)	03.00
Vogt P.	Bussmann-Holder	99	51	–	98	(83)	77.00
Wagner H.	Wagner H.	85	151	–	221	(79)	73.00
Walker M.	Stewart J.	69	69	–	115	(73)	97.60
Wanders G.	Wanders G.	57	22	–	40	(71)	13.75
Weber H.J.	Arenhövel H.	65	58	–	91	(72)	21.00
Weinstein M.	Weisntein M.	60	32	–	73	(71)	11.40
Wess J.	Wess J.	50	132	–	142	(69)	11.30
Wess J.	Wess J.	60	1	–	17	(71)	11.30
Wick G.C.	Wick G.C.	108	275	–	278	(86)	29.00
Wiik B.H.	Wiik B.H.	86	1	–	262	(79)	12.00
Wildermuth K.	Wildermuth K.	41	1	–	172	(66)	21.00
Wilson R.	Wilson R.	39	43	–	54	(65)	13.40
Wissmann P.	Wissmann P.	77	1	–	96	(75)	73.00
Wolf G.	Wolf G.	59	77	–	113	(71)	13.60
Wolf G.	Wiik B.H.	86	1	–	262	(79)	12.00
Yang C.N.	Yang C.N.	108	279	–	283	(83)	29.00
Zeitnitz B.	Fries D.	100	1	–	223	(82)	21.00
Zeller R.	Dederichs P.H.	87	1	–	170	(80)	61.00
Zeppenfeld K.	Daniels J.	54	77	–	135	(70)	71.00
Zimmermann W.	Zimmermann W.	50	143	–	156	(69)	11.00
Zinn-Justin J.	Zinn-Justin J.	57	248	–	270	(71)	13.75

Printing: COLOR-DRUCK DORFI GmbH, Berlin
Binding: Buchbinderei Lüderitz & Bauer, Berlin

Springer Tracts in Modern Physics

Editors: G. Höhler, E.A. Niekisch

Springer-Verlag
Berlin
Heidelberg
New York
London
Paris
Tokyo
Hong Kong
Barcelona

Springer Tracts in Modern Physics

* denotes a volume which contains a Classified Index starting from Volume 36